ENVIRONMENTAL HEALTH - PHYSICAL, CHEMICAL AND BIOLOGICAL FACTORS

CARBON MONOXIDE

SOURCES, USES AND HAZARDS

Environmental Health - Physical, Chemical and Biological Factors

Additional books in this series can be found on Nova's website under the Series tab.

Additional E-books in this series can be found on Nova's website under the E-book tab.

Public Health in the 21st Century

Additional books in this series can be found on Nova's website under the Series tab.

Additional E-books in this series can be found on Nova's website under the E-book tab.

ENVIRONMENTAL HEALTH - PHYSICAL, CHEMICAL AND BIOLOGICAL FACTORS

CARBON MONOXIDE

SOURCES, USES AND HAZARDS

DANTE DILORETO
AND
IDINA CORCORAN
EDITORS

Nova Science Publishers, Inc.
New York

For permission to use material from this book please contact us:
Telephone 631-231-7269; Fax 631-231-8175
Web Site: http://www.novapublishers.com

LIBRARY OF CONGRESS CATALOGING-IN-PUBLICATION DATA

Carbon monoxide : sources, uses, and hazards / editors, Dante DiLoreto and Idina Corcoran.
p. cm.
Includes index.
ISBN 978-1-61942-055-7 (hardcover)
1. Carbon monoxide. I. DiLoreto, Dante. II. Corcoran, Idina.
QD181.C1C325 2011
546'.6812--dc23

2011040799

Published by Nova Science Publishers, Inc. † New York

CONTENTS

PREFACE

Carbon monoxide (CO) is a colorless, odorless, highly toxic, poisonous gas. Exposure to a few hundreds ppm can cause permanent health damages or death. CO can thus be considered as one of the most dangerous environmental pollutants and therefore, its removal and detection is of extreme importance for safe living. This book presents current research in the study of the sources, uses and hazards of carbon monoxide. Topics discussed include the cardiac effects of carbon monoxide intoxication; CO oxidation using gold supported on Ce-Mn-O composite materials; heterogeneous catalysis in carbonylative coupling reactions and the evaluation of carboxyhemoglobin (CO-Hb) by oximeter in the forensic practice.

Chapter 1 - Carbon monoxide (CO) toxicity has been well known for a long time however the exact mechanism through which it damages the body has been revisited. The poor correlation with carboxyhemoglobin levels led to attribute CO toxicity also to the high affinity binding with myoglobin, to the competitive inhibition of cytochrome oxidase and to the induction of inflammation and oxidative stress. The discovery that carbon monoxide is not only a toxic pollutant but also a substance normally produced at very low doses in the body where it exercises a signalling effect in a myriad of functions like the regulation of the vascular tone, the mitochondrial biogenesis and the inflammation, aroused much interest and has even contributed to clarify the mechanisms of its toxic action. The competitive inhibition of the cytochrome oxidase, shared by other gaseous transmitters (gasotransmitters), of similar chemical structure, like Hydrogen Sulphide and Nitric Oxide, is important not only as a cause of toxicity but also for the regulation of cellular oxygen consumption and for the involvement in the pathophysiological mechanism of many diseases. The clinical picture may be very variable so that CO poisoning has been named "the great imitator". The most common cardiac manifestations are ischemic chest pain, arrhythmias and hypotension. Carbon monoxide can injury particularly the damaged cardiovascular system but also the healthy man is subjected to its toxicity, whose spectrum ranges from myocardial infarction to prolonged ischemia showing frequently itself in the clinical setting with a picture of stunned myocardium. CO intoxication has been showed to substantially increase long-term mortality. The clinical assessment is based on the evaluation of ECG, echocardiography and of the biochemical markers (troponin and BNP). A quick removal from the toxic environment and oxygen therapy are the basis of treatment while it is yet discussed the role played by hyperbaric oxygen therapy in the setting of CO poisoning in terms of increased survival or improved long-term outcomes.

Chapter 2 - Carbon monoxide (CO) intoxication is one of the main causes of mortality in the world because of its tastelessness, colorlessness, and odorlessness without irritating properties, leading to unawareness of excessive inhalation of potentially lethal concentrations for victims. Moreover, the frequency of non-lethal CO intoxications is even higher but difficult to measure for the nonspecific presentations caused by this intoxicant. In addition, CO measurements at the time of actually reaching the emergency unit may not always correlate with symptoms, findings, and prognosis in acute CO intoxications. Although non-invasive and quick brain imaging, including computerized tomography (CT) and magnetic resonance imaging (MRI), may help clinicians to approach typical patients with CO intoxication, however, they are not always sensitive to patients with varied extent of CO inhalation. Given the universal use of fossil energy sources, all of the individuals are dangerously threatened by this "invisible killer". Understanding the causes, risks, and prevention of CO intoxication may decrease needless mortality and morbidity.

Chapter 3 - Carbon monoxide (CO) is a colourless, odourless, highly toxic, poisonous gas. Exposure to a few hundreds ppm can cause permanent health damages or even death. Low level exposure, if continued, can also cause health problems in the long term. CO can thus be considered as one of the most dangerous environmental pollutants. Therefore, its removal and detection is of extreme importance for safe living. Catalysts which promote CO oxidation at room temperature will find many practical applications that include air-purification devices for indoor space (including gas masks), and for cleaning automotive exhaust gases. CO removal is also required in CO_2 lasers, and in polymer electrolyte fuel cell (PEFC) usage, where the CO concentration in the hydrogen rich gas must be selectively reduced to less than 100 ppm.

Haruta and co-workers have intensively investigated the ability of gold catalysts to catalyse this reaction. In their early work in the 80s, the activity of supported gold catalysts came as a surprise, but it is now well known that supported gold nanocatalysts are effective for CO oxidation at very low temperatures, being active even at -70°C. However, activities depend on the choice of the metal oxide support. In the present work, Ce-Mn-O supports with different Ce/Mn ratios were prepared by co-precipitation, in addition to cerium oxide only and manganese oxide only. These supports were loaded with gold by a double impregnation method. Samples were characterised by adsorption of N_2 at -196°C, temperature programmed reduction (TPR), X-ray diffraction (XRD), high-resolution transmission electron microscopy (HRTEM) and high-angle annular dark-field imaging (HAADF) with Z-contrast. CO oxidation was studied using a continuous-flow reactor connected to a gas chromatograph. It was found that the higher the amount of ceria in the sample, the most active it was for CO oxidation. The ability to convert CO seems to be independent of the surface area of the supports, but seems to vary according to the temperature of the first reduction peak of the TPR spectrum, in the case of samples loaded with gold. Also it varies according to the gold nanoparticle size.

Chapter 4 - Carbon monoxide (CO) in combination with transition metal catalysts has become a versatile reagent in organic synthesis. Insertion of the CO molecule introduces a carbonyl group, which is one of the most common functionalities in bioactive/industrially important compounds. The bulk chemical industry also takes advantage of CO as one of the cost efficient C_1 building block for various transformations like olefins into aldehydes, carboxylic acid derivatives, or alcohols by way of carbonylation reactions, thus providing a straight forward access to valuable intermediates for the synthesis of polymers and other consumables. After the first reports by Heck and co-workers in the 1970s on palladium catalyzed carbonylation reactions, the scope of CO chemistry has expanded considerably. Since then various new classes of carbonylation reactions such as alkoxycarbonylation, phenoxycarbonylation, aminocarbonylation, thiocarbonylation and carbonylative coupling reactions like carbonylative Suzuki coupling, carbonylative Sonogashira coupling, carbonylative Heck coupling etc. have been emerged and mainly using palladium as a catalyst of choice with good catalytic activity and selectivity. Several research groups have developed various homogeneous palladium complexes based catalytic systems for these carbonylation reactions in which catalyst are in dissolved with the substrate. Despite of their potential utility, homogeneous catalytic systems have one or more drawback such as difficulties in catalyst and product separation, loss of expensive metal and non-recyclability of metal catalyst limiting their applications. Therefore, to craft synthetic protocol more economical and efficient, heterogeneous catalysis was found to be the best alternative to homogeneous processes. Also the catalyst recovery decreases the contamination of desired products with trace metal or ligand impurities. The contamination of catalyst even in ppm concentration is not desirable for pharmaceutical applications. Herein, we summarize the recent developments in use of heterogeneous catalysts for these new classes of carbonylative coupling reactions.

Chapter 5 - Carbon monoxide (CO) is the leading cause of poisoning death in Japan. As CO is generated by the imperfect combustion of various organic compounds, fire and other causes, measurement of the saturation of carboxyhemoglobin (CO-Hb) in blood is important in forensic practice. Quantitation of CO-Hb is usually performed by various methods such as gas chromatography or spectrophotometry. In the present study, the authors determined CO-Hb saturation in blood samples obtained from fire-related deaths and CO poisoning cases using a portable oximeter AVOX 4000 (AVOX). The CO-Hb was also measured by the conventional spectrophotometric method. There was a good correlation between the values obtained by AVOX and by the conventional method, within the range of 10-75% of CO-Hb. In over 75% of CO-Hb by the conventional method, the values were indicated "75%<" by AVOX. AVOX is suitable for forensic practice as a new tool by easy handling with a small sample in comparison with conventional methods.

Chapter 6 - Between 1992 and 2004, 1,381 people living in Ciudad Juárez, México, were intoxicated, and 167 more have lost their lives to carbon-monoxide poisoning. This study examines 91 incidents of carbon-monoxide intoxication from October to February 2003-2004. Gas heaters were identified as the cause of 59% of the incidents. Carbon-monoxide intoxication incidents are associated with colder temperatures on both a monthly average and a daily basis. A neighborhood-level analysis suggests that households in the middle of the income spectrum are disproportionately at risk. An exploratory comparison of risk factors between the case data and baseline data from three neighborhoods found that gas heaters are associated with higher risks than wood or charcoal heaters. In 32% of the intoxication incidents, doors and windows of the home were sealed with plastic, newspaper or other materials. While differences are noted in the relative risks for heaters produced by different manufacturers, these differences are likely not applicable to current models as the manufacturers have introduced new designs. Interventions supported by this research include efforts to encourage proper ventilation of homes, the replacement of older heaters with newer models that have oxygen depletion sensors, the use of carbon monoxide detectors, and turning off heaters while sleeping.

In: Carbon Monoxide: Sources, Uses and Hazards ISBN:978-1-61942-055-7
Editors: D. DiLoreto et al.

Chapter 1

THE CARDIAC EFFECTS OF CARBON MONOXIDE INTOXICATION

Luca Siracusano* and Viviana Girasole
Department of Neuroscience, Psychiatric and Anesthesiological Sciences,
University of Messina, School of Medicine,
Policlinico Universitario G. Martino, Messina, Italy

ABSTRACT

Carbon monoxide (CO) toxicity has been well known for a long time however the exact mechanism through which it damages the body has been revisited.

The poor correlation with carboxyhemoglobin levels led to attribute CO toxicity also to the high affinity binding with myoglobin, to the competitive inhibition of cytochrome oxidase and to the induction of inflammation and oxidative stress.

The discovery that carbon monoxide is not only a toxic pollutant but also a substance normally produced at very low doses in the body where it exercises a signalling effect in a myriad of functions like the regulation of the vascular tone, the mitochondrial biogenesis and the inflammation, aroused much interest and has even contributed to clarify the mechanisms of its toxic action.

---

* Luca Siracusano viale Regina Margherita 28 98121 Messina Italy 0039 090 46461 Fax 0039 090 2212494 e-mail lsiracusano@unime.it1.

The competitive inhibition of the cytochrome oxidase, shared by other gaseous transmitters (gasotransmitters), of similar chemical structure, like Hydrogen Sulphide and Nitric Oxide, is important not only as a cause of toxicity but also for the regulation of cellular oxygen consumption and for the involvement in the pathophysiological mechanism of many diseases.

The clinical picture may be very variable so that CO poisoning has been named "the great imitator". The most common cardiac manifestations are ischemic chest pain, arrhythmias and hypotension. Carbon monoxide can injury particularly the damaged cardiovascular system but also the healthy man is subjected to its toxicity, whose spectrum ranges from myocardial infarction to prolonged ischemia showing frequently itself in the clinical setting with a picture of stunned myocardium. CO intoxication has been showed to substantially increase long-term mortality. The clinical assessment is based on the evaluation of ECG, echocardiography and of the biochemical markers (troponin and BNP).

A quick removal from the toxic environment and oxygen therapy are the basis of treatment while it is yet discussed the role played by hyperbaric oxygen therapy in the setting of CO poisoning in terms of increased survival or improved long-term outcomes.

Carbon monoxide (CO) poisoning is today the most common cause of morbidity and mortality for intoxication (50.000emergence department visits and 3.500 deaths per year in the United States) (1) and it is frequently difficult to recognize for the nonspecificity of the symptoms (2); this is made easy by the chemico-physical characteristics of this gas. These characteristic of CO deserved it the name of "silent killer", whereas the multifaceted clinical picture gained it the name of "the great imitator". The main target of carbon monoxide poisoning are the brain and the heart for their high oxygen consumption; the two organs show frequently a different level of impairment probably because the toxicity is not caused by a single mechanism and they present some molecular targets in different amounts. The ways in which carbon monoxide regulates the organ functions, particularly the cardiac one, physiologically and compromises them in toxic doses and the related clinical manifestations are the subject of this chapter.

Chemico-Physical Properties of Carbon Monoxide

CO is a colourless, odourless and tasteless gas, not irritating, not suffocating, inflammable at high concentration. CO is the by-product of the incomplete combustion of combustible hydrocarbons (fossil, fuels, oil, gas, wood, coal, tobacco etc) caused by insufficient oxygen supply for a complete oxidation to carbon dioxide (CO_2). The most common causes of poisoning are the heating and cooking devices burning these substances, central heating burning gases, fireplaces or car exhaust gases in poorly ventilated areas even though the use of catalytic converters has reduced CO emissions from 3,5 to 0,5% (3) . Forest fires or volcanic eruptions may also be causes of intoxication. The average carbon monoxide in the natural atmosphere is 0,1 ppm, the levels in home are 0,5-5 ppm. but the presence of poorly adjusted stoves may increase the level to over 30 ppm. The recommended dose in the environment by OMS is 10 ppm; doses around 200 ppm cause headache after 2-3 hours and above 2000 ppm cause death after 1 h (3). Heavy smokers may develop a chronic intoxication reaching a level of 10-15% of carboxyhemoglobin (CO-Hb). CO-Hb levels between 15 and 20% seem to be well tolerated in humans and represent the threshold value beyond which the signs of poisoning begin in previously healthy persons.

Pathophysiological Mechanism of CO Toxicity

The damage by CO depends on the relative concentration of CO and oxygen in the environment, minute ventilation, the duration of exposure, metabolic rate and hemoglobin concentration (4). Carbon monoxide is only slightly metabolized in the body [only 1% is transformed in CO2 by cytochrome C oxidase (COX)], it binds to proteins and only less than 1% is in solution but this small amount has a very important role in the dynamics of cell damage. In the body CO binds almost exclusively to the transition metals present in the prosthetic group of many proteins modifying their biological activity; the most relevant example for its toxicity is the interaction with proteins containing iron or other transition metals such as hemoglobin, myoglobin (Mb), cytochromes and soluble guanylate cyclase (sGC) (5). Overall 10-15% of CO is bound to proteins.

The binding of CO to a protein is regulated by the partition coefficient of Warburg (K) that depends on the concentrations of both CO and O_2 at which the protein is half saturated with CO; a reduction in PO_2, such as that induced by local ischemia, will thereby increase CO toxicity (6,7).

CO avidly binds (with an affinity 200-250 times superior to oxygen) the ferrous heme moiety [but contrary to Nitric Oxide (NO) not the ferric iron] leading to the formation of CO-Hb so reducing oxygen transport and causing hypoxia; the binding with Hb is much stronger than with oxygen and slow-releasing. Furthermore when Hb binds CO, the non CO-bound molecules of Hb increase their affinity for oxygen, due to an allosteric effect of CO, so causing a shift to the left of the oxyhemoglobin (Hb-O_2) dissociation curve that changes its shape from sigmoid to hyperbolic with an overall reduction in oxygen delivery to the tissues (8). Tissue PO_2 is near to venous tissue PO_2 and may be as low as 5-30 mm Hg in organs such as heart, liver and brain (6), in this last watershed zones like pallidus are the most frequently damaged. Hypoxia alone however cannot justify the damage by CO according to the Haldane theory (9) of pure hypoxia as showed by the elegant experiment by Goldbaum (10) who observed that transfusing partially exanguinated dogs with donor blood containing 80% CO-Hb, so providing an average carboxyhemoglobin level of 60%, did not produce CO toxicity, whereas inhalation of CO (13 percent in air) for 15 minutes caused death within 15 to 65 minutes. The amount of non Hb-bound CO, although small, is critical in further increasing organ damage, inhibiting cell respiration by binding to COX and to Mb. CO binds to Mb in skeletal and cardiac muscle (6) with an affinity 40 times superior than oxygen and 4 times greater for cardiac than for skeletal muscle myoglobin; CO binding to myoglobin is further increased during hypoxia (7). As carboxymyoglobin (COMb) dissociation is slower than that of CO-Hb, Mb represents a significant storage site for CO (11) and the late increase in CO release, one of the mechanisms responsible for the delayed neurologic damage, can be ascribed to the dissociation of the binding with myoglobin, slower than the binding to Hb (7). Mb has many recognized functions in myocardium and skeletal muscle where is abundantly present (100 to 400 μM\Kg) conferring the characteristic red colour: it can store oxygen to buffer short ischemic events or also the physiological increase in oxygen consumption during systole or exercise (12). According to the "facilitated oxygen diffusion" theory the mobile molecule of Mb may transport oxygen from plasma membrane to the mitochondria (13). Moreover Mb has a scavenger role for reactive oxygen species (ROS) and NO, so regulating the depressing effect of NO on mitochondrial respiration (14,15).

The importance of myoglobin in cellular oxygenation has been assessed in myoglobin deficient mice that show a depressed cardiac function only in hypoxic conditions with a 30% reduction in systolic function reversible on restoring normoxia or with a nonspecific Nitric Oxide Synthase (NOS) inhibitor, showing that myocardial depression in this setting is NO-mediated (16). Whereas the negative effects of CO-Mb formation are well known less recognized are the consequences of CO binding to Mb on its scavenging function of NO. Wunderlich observed in the isolated heart of iNOS-overexpressing mice that the administration of CO 20% resulted in a marked inhibition of the scavenging role of Mb on NO that accumulated in the myocardiocytes with a marked impairment of cardiac contractile force reversible by administration of a NOS inhibitor (17); the mechanisms depressing cardiac performances by CO binding to Mb are therefore manifold. Also in the brain is present a protein similar to myoglobin, the neuroglobin, that plays an important role in the protection of brain neurons from ischemic and hypoxic injuries, that upregulate its production.

It was at first considered as a protein devoted to oxygen storage during hypoxia and ischemia but its concentration far smaller (1μM\Kg) such as its high P_{50} make unlikely a function in oxygen transport and storage except in retina where it has a much higher concentration (100μM\Kg); neuroglobin is now considered important in NO and ROS scavenging, oxygen sensing and signal transduction (18,19). CO has a very high affinity for the Fe^{2+} in hemea_3 of the COX so inhibiting respiration despite it has minor affinity than oxygen for COX (20,21); the binding however, when established, is slowly releasing. The inhibition of the mitochondrial electron transport increases the release of superoxide at the Complex III level and the production by superoxide-dysmutase (SOD) of hydrogen peroxide (H_2O_2); the low, physiologic levels of H_2O_2 activate cellular signalling by CO (22), whereas the high toxic levels of CO cause oxidative stress. The inhibition of cellular respiration by CO has been demonstrated in vitro but the ability of the physiological levels to depress COX in vivo is uncertain, it is possible however, as suggested by Moncada and his group (23), that the high level of CO (such as observed during hypoxia, inflammation, sepsis) may actually modulate oxygen consumption in vivo. Moncada suggested that also the depressing effect of NO on COX activity, absent in vivo because of the scavenging effect of Mb on NO, becomes probably important in inflammatory pathologies (24) .

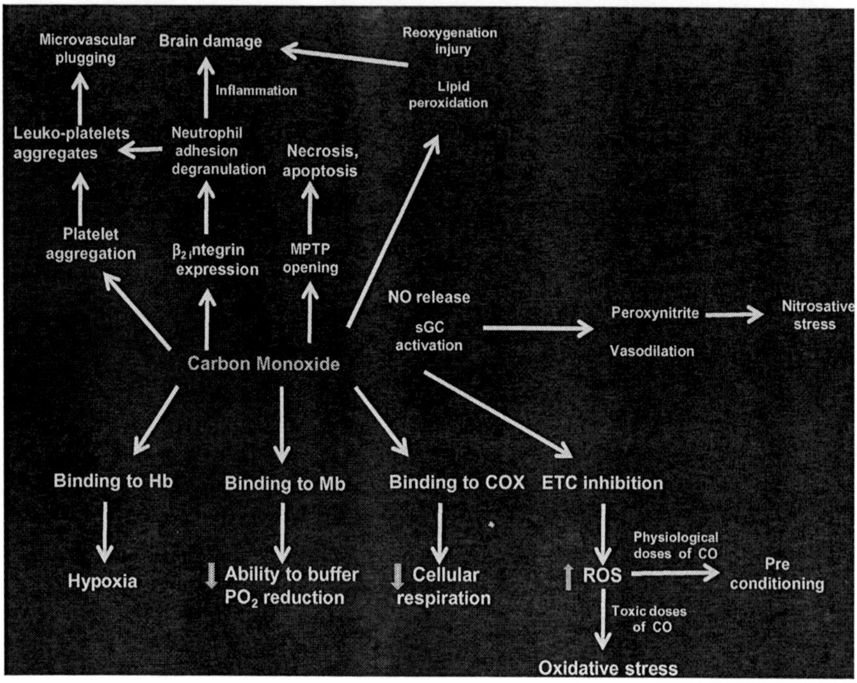

Figure 1. Pathophysiological mechanisms of CO toxicity.

Other gaseous molecules share with CO the ability to interact with COX, to inhibit it competitively and to regulate at this level the body oxygen consumption (25,26); they are called gasotransmitters and include apart from CO also NO and Hydrogen Sulphide (H_2S), this last molecule was first shown able to induce experimentally in mice, but not in pigs and superior mammals, a particular state of depression of the metabolism and of the resistance to hypoxia, called "suspended animation" (27), a similar observation has been made with NO and CO in Drosophila (28,29). Exploiting this property of gasotransmitters in the treatment of myocardial and cerebral hypoxia\ischemia is tempting, considered also the limitations of the current treatment with hypothermia of the hypoxic organ injury even though they also have a high toxicity (30). Although difficult to utilize at present, metabolic depression, similar to hibernation, mediated by depression of the COX activity might be in the future a strategy to reduce organ failure after trauma and hypoxia. A similar state of depressed metabolism may also be induced clinically and experimentally by excess NO during sepsis (metabolic hibernation) (31-33). Interestingly also the levels of CO and H_2S are increased during sepsis and in other diseases (34-36). CO is slowly removed from tissues through transformation into CO-Hb that behaves as a buffer inhibiting the cession to the tissues, the outcome is therefore more linked to the duration of the exposition and to the plasma-dissolved CO than to the concentration in the environment. Vreman et al (37) studied the distribution in the tissues of CO in poisoned patients, showing the highest concentration in blood, spleen, lung, kidney, heart and muscle whereas brain and adipose tissues contained the lowest amounts of CO probably because of the low CO solubility in lipids and the low content of heme compounds, on the other hand the content in blood of different organs may affect the CO levels they contain. The importance of COX binding in CO toxicity is showed by the demonstration that the activity of this enzyme returns to the normal level more slowly than CO-Hb after treatment (38).

The damage induced by CO poisoning is also dependent on on the reoxygenation injury in the brain inducing necrosis and apoptosis (39), on the oxidative stress caused by the interference with the electron transport chain (ETC) and by the increase in the levels of free heme (40). The neuronal injury is amplified by the lipid peroxidation and by the degradation of myelin basic protein in the brain (41,42). Toxic levels of CO have been shown to cause β2-integrin-dependent adherence of neutrophils to the injured endothelium in a rat model (43). The precise mechanism of leucocytes adherence to endothelium is complex and is mediated by the ability of CO to react with both platelets and neutrophils activating intravascularly these last cells. While low levels of CO inhibit platelet aggregation through interaction with sGC, increase in cGMP and reduction in intracellular Ca^{2+} protecting against organ damage (44), the toxic levels have been shown to bind to platelets hemoproteins displacing NO and increasing its release; this phenomenon by which a substance may reverse its biological effects beyond a critical concentration is called hormesis (38). In this setting of oxidative stress, NO released by platelets reacts with ROS produced by neutrophils giving origin to peroxynitrite; this highly toxic molecule activates platelets, causes expression of adhesion molecules with neutrophil degranulation and formation of platelet–neutrophil aggregates (43). The interaction between platelets and neutrophils may thus damage tissues in many ways through microvascular plugging, oxidative stress and inflammation.

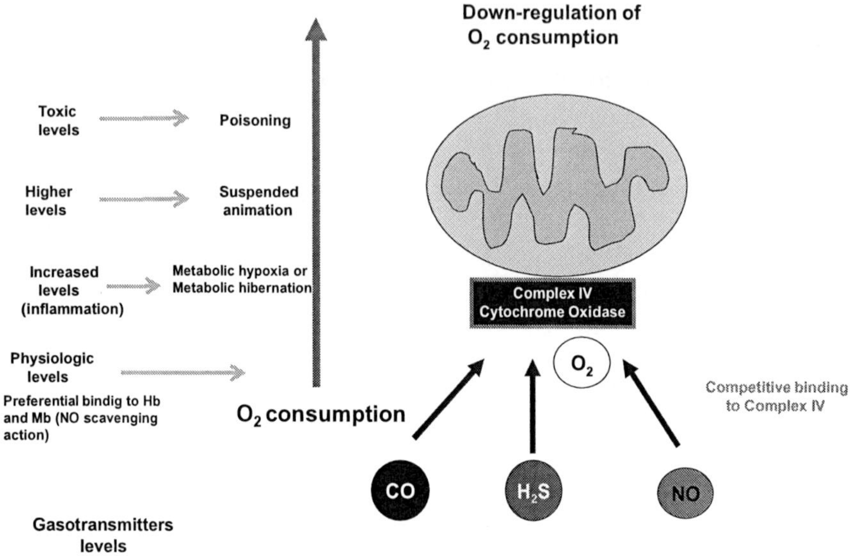

Figure 2.The picture shows that increasing levels of the 3 gasotransmitters give rise to difference pathophysiological effects.

Biological Effects of Endogenous CO

The catabolism of heme by the enzymes heme-oxygenase 1 and 2 (HO-1 and 2) gives rise to equimolecular amounts of biliverdin, quickly oxidized to bilirubin, iron and CO provided with many important signalling and protective functions. Whereas HO-2 and the less studied HO-3 are the constitutive forms, HO-1 is the inducible form of the enzyme produced in response to various types of challenges increasing oxidative stress (hypoxia, cytokines, heavy metals etc)(45). HO-1 is localized in the ER, plasma membrane caveolae, mitochondria and nucleus. The preferential localization is perhaps in the ER because many conditions which induces this enzyme are also able to induce ER stress, that is a damage to ER caused by an accumulation of unfolded proteins in pathological conditions such as nutrient excess and obesity (46). To function HO needs also NADP(H), cytochrome p450 reductase (CPR) and biliverdin reductase (BVR). The protective role of HO-1 is demonstrated by the first described case of HO-1 deficiency in humans, a boy with growth retardation, severe persistent endothelial damage and early vascular atherosclerotic alterations, abnormal coagulation and fibrinolysis and persistent haemolytic anemia (47). HO-1-2 give rise to biliverdin and bilirubin provided with antioxidant properties and reduce the level of ROS degrading heme and so reducing the toxicity of its ferrous molecule mediated by hydroxyl formation (48). The majority of the effects attributed to HO isoenzymes are however mediated by CO (49-50), overproduced because of the induction of HO-1 by the stressing events. A less important HO-indipendent CO production (51) is also present.

The low endogenous levels of CO have been involved in a myriad of physiological functions:

1) Although it is not itself an antioxidant CO activates antioxidant genes and reduces apoptosis (45): low levels of CO induce a mild burst of ROS thus inducing antioxidant enzymes (Catalase, Manganese Super Oxydo Dysmutase, MnSOD) and glutathione.
2) CO has been shown to precondition cardiomyogenic cells (52) through activation of PI3K\Akt protecting them against ischemia-reperfusion injury (IRI) (53). This "early preconditioning effect" is mediated by the production of ROS in the respiratory chain which activates Akt with opening of mitoKATP and prevention of the opening of the mitochondrial permeability pore transition (MPTP) (a pore opening in the internal mitochondrial membrane when ROS and Ca^{2+} accumulate that allows passage of solutes of a size up to 1500 Da causing the rupture of the external mitochondrial membrane and mitochondrial death), event that may lead to cell death by necrosis and apoptosis (54); on the contrary the high toxic levels of CO may induce the opening of MPTP (55). ROS production induced by CO is also involved in the stabilization of hypoxia-inducible factor-1 (HIF-1) and the consequent cardioprotection that has been shown to be mediated by HO-1 activation (9,56,57).

   HIF-1 is activated during CO poisoning to regulate mitochondrial ROS production enhanced under hypoxic and ischemic conditions. HIF-1 induces a mitochondrial reprogramming that optimizes the efficiency of electron transport and minimize O_2^- production in hypoxic conditions (58).
3) Akt activation may also phosphorylate many target enzymes such as nuclear respiratory factor 1 (NRF1) and NRF2, peroxisome proliferator-activated receptor gamma coactivator-1 (PGC-1) and mitochondrial transcription factor A (TFAM) so leading to an increased production of mitochondrial DNA (mtDNA) copy number,

to increased mitochondrial biogenesis and to protection against apoptosis (59). Lancel observed that low doses of CO, released by Metal carbonyl-based compounds [CO-releasing molecules (CO-RMs)], may rescue mice by lethal sepsis increasing the expression of PGC-1α and improving mitochondrial function (60). As APN, adipokine sharing many effects with CO, is depleted in sepsis worsening outcome through proinflammatory cytokines production (61) the existence of an axis has been proposed between HO-1 and APN, modulating in sepsis not only the inflammatory response but also the mitochondrial biogenesis through an increase in PGC-1 production (62).

4) CO activates sGC increasing intracellular cGMP, activating high-conductance calcium-activated K channels (bK_{Ca}) and inhibiting the P-450 system (63-64), thus inhibiting the vasoconstrictor effect of its metabolite 20-HETE; moreover CO may release the more powerful sGC activator NO (65). All these effects may explain the vasodilation by CO and the first symptoms of intoxication: mild headache by cerebral vasodilation and cutaneous vessels dilatation.
5) CO has been experimentally showed to inhibit LPS-induced pro-inflammatory interleukins secretion increasing on the contrary the contra-inflammatory IL-10 (66) so modulating inflammation. This effect is mediated by the regulatory role on MAPK function, with activation of p38 MAPK and downregulation of ERK1\2 (67) probably through protein phosphatase 2C (PP2C), a serine/threonine phosphatase that is essential for regulating cellular stress responses and containing Mn^{++} in its active site (5). MAPK modulation exerts anti-inflammatory and antiapoptotic actions and may contribute to reduce IRI (53). CO, inhibiting ERK1\2, reduces also the expression of EGR-1, potent transcriptional activator of deleterious thrombotic and inflammatory cascades in a rat lung ischemic injury model (68).
6) Recently was emphasized the ability of CO to interact with ion channels either directly or through the interference with transition-metal containing proteins and particularly heme that has been shown to be an allosteric regulator of human maxi-K^{+} channels (5). Myocardial cells express HO-2 and, under stress conditions, also HO-1 that are known to protect heart against IRI through the release of CO: HO-1 ko mice present an increased infarct size whereas OH-1 overexpression reduces it compared to wild type mice. The inhibition of Ca^{2+} influx through cardiac L-type Ca^{2+} channels may also play a

role in the protection conferred by CO (69). Moreover CO was demonstrated to inhibit epithelial Na^+ channel (ENAC), delayed rectifier K+ channels Kv2.1 and, as already seen, high-conductance bK_{Ca} calcium-regulated potassium channels modifying the vascular function (for reviews 70,71)

The anti-inflammatory, vasodilating and antiischemic properties CO possesses have been recently utilized by administration of the already mentioned CO-RMs, able to release controlled amounts of CO in the body; their protective effect is similar to that conferred by HO1 overexpression (72). CO therapeutic administration has found a field of use, so far only experimentally, in many different indications ranging from acute lung injury, organ transplantation, inflammatory diseases and atherosclerosis to sepsis and organ damage (73,74).

Clinical Manifestation of CO Poisoning

CO poisoning has no pathognomonic symptoms and the first manifestations are vague and nonspecific, frequently with slight headache by dilatation of brain vessels, visual troubles, dizziness, nausea, vomiting, general malaise, simulating flu-like symptoms in the mild forms (3,4,6); neuropsychological impairment with deterioration in memory and cognitive functions may be present also at low level of CO inhalation (17-100 ppm) (75). In the moderate poisoning may be present chest pain, tachycardia and tachypnea, rhabdomyolysis (caused by binding of CO to muscle Mb), neuropsychological impairment and altered state of consciousness, anxiety or depression. The most severe cases may present with seizures, lethargy, coma, cardiogenic and noncardiogenic pulmonary edema and sudden arrhythmic death (3).The classic appearance with red cherry lips and cyanosis is less frequent. A worsening may be present of pre-existing diseases like chronic broncopulmonary disease (COPD), cardiac diseases, posttraumatic syndrome (2).

Cerebral damage is probably mediated by a process of demyelination of the cerebral white matter as showed by magnetic resonance imaging (MRI) and computed tomography (CT) scans (6,76,77). The heart of poisoned p. presents on autopsy scattered punctiform hemorrhages throughout the cardiac walls (78); myocardial fibrosis has also been detected by magnetic resonance imaging (MRI) in severe poisoning (79). Delayed neuropsychological impairment, one of the most feared complications, presenting as cognitive or affective disorders, hearing loss, motor disturbances, psychosis or dementia, may be detected in 2-40% of the p. in the following 2-6 weeks (6,76). The pathological lesions underlying these symptoms are demyelination, petechiae, oedema and necrosis. MRI can show T2-weighted hyperintensities, lesions of the basal ganglia and atrophy of hippocampi and pallidus, this last being the most frequently damaged (6,76).

The severity of poisoning is mainly related to the amount of CO dissolved in the plasma; whereas CO-Hb in blood correlates well with the dose of carbon monoxide the patient has been breathing and represents thus an excellent biomarker for recent carbon monoxide exposure if an equilibrium had already been achieved, the clinical picture instead badly related to it if the plasma levels are measured out of the scene of poisoning and increasingly as time elapses and treatment is established. Indeed oxygen administration may alter CO-Hb level and the concentration of CO dissolved in the plasma may be responsible for the binding to COX and Mb, so the damage can go on due to the developing oxidative stress and to impaired cellular respiration. CO poisoning is more dangerous in the pregnant woman because, although pregnancy does not increase the severity of outcome for the mother, the fetus is more susceptible to its toxic action because of the dissociation curve of Hb-O_2 is shifted to the left. Fetus PO_2 is 25% than that of the mother and metabolism of CO-Hb is slower so that there is no correlation between the plasma level of CO-Hb of the fetus and of the mother (2-4,80-82). Remains yet doubtful if CO has a teratogenic effect (83).

The diagnosis is based on the history: the exposition to car exhaust, stoves, fireplaces in restricted environment, the presence of other components of the family or coworkers with similar symptoms should induce in suspect (84). The data provided by pulse oximetry could result normal because CO-HB is often not differentiated from Hb-O_2 so direct measurements of both CO-Hb and Hb-O_2 levels are necessary although they are more useful for diagnosing CO intoxication than to predict its seriousness. Arterial blood gas analysis for the same reason may give normal PO_2 levels, it may also show a metabolic acidosis caused by anaerobic metabolism due to hypoxia. Hampson and Hauff observed a significant correlation between initial CO-Hb and the decrease in the initial pH (85) although it was impossible to predict CO-Hb only from pH, chiefly when the levels of the former are low. These data confirm that hypoxia is not the only mechanism of CO-induced damage and underline the importance of the cardiac impairment on pH values and on outcome. In nonsmokers the normal CO-Hb, expression of the low level physiological CO production, is 1% and values above 2% may already be expression of poisoning (85), heavy smokers can have a CO-Hb level up to 10-15% (80), although normally in smokers a value above 9% is considered suspect (85). Values over 20% represent the threshold of poisoning but in diseased subjects symptoms may appear also with amount of 3-5% (86).

CARDIOVASCULAR TOXICITY OF CO

The toxicity for the heart is distinguished by the importance of CO binding to Mb present in high concentration in the heart and with the greatest importance for cardiac function. Moreover the hemodynamic and coronary effects of CO have a role much more relevant than in other organs. Although the hemodynamic response to CO in the clinical and experimental settings varies in relation to dose, rate of delivery, duration, state of consciousness, and level of exertion generally there is a tendency, in less severe poisoning, to an increase in heart rate (HR) and in cardiac index (CI) to compensate for reduced oxygen delivery; higher doses however depress both HR and CI (87). A recent study in a paced and perfused model of isolated rat hearts submitted to a variable concentration of the mixture O_2-CO has further stressed the effects of the direct myocardial toxicity of this gas showing a depressed cardiac function not attributable to CO-Hb because Hb was not present in the experimental environment. When 30% CO in oxygen has been used the heart has shown a significant depression but has been able to recover, whereas this has not been possible when CO was raised to 40% (88).

Andre and colleagues (89) to reproduce the effects of chronic exposure to CO present in pollution, exposed for 4 weeks rats to filtered air (CO 1 ppm) or air enriched with CO (30 ppm with five peaks of 100 ppm per 24-h period). They showed the induction of interstitial and perivascular fibrosis and an altered systolic and diastolic function. The depressed contraction of single rat ventricular myocytes was related to the reduction of both systolic Ca^{2+} release and of myofilament Ca^{2+} sensitivity. The reduced sarco/endoplasmic reticulum Ca^{2+}-ATPase 2a (SERCA-2a) expression induced by CO exposure accounted likely for the impairment in Ca^{2+} reuptake and the delay in cell relaxation. The ventricular arrhythmias (VAs) observed in this study are mediated by Ca^{2+} overload consistent with chronic β-adrenergic overstimulation during CO exposure

The effect on the coronaries depends on the vascular effects of CO. As already underlined CO has a vasodilatory effects through many possible mechanisms (63,64): evidences are however present that in some circumstances CO may behave as a vasoconstrictor suppressing endothelial NO synthase activity and exerting endothelium-dependent vasoconstrictive influence in isolated arteries (64,90,91). Favory and colleagues studied in an ex vivo isolated heart rat preparation the effects of moderate CO exposure (200 ppm with mean COHb level of 11%) on cardiac contractile and coronary functions (92). In this model CO induced an increase in LV contractility not prevented by β-block and attributable to increased Ca^{2+} sensitivity of myofilaments. Coronary function was modified with reduced coronary blood flow reserve and abnormalities of endothelium-dependent and -independent vasorelaxation associated to a decreased cGMP/cAMP ratio. The AA observe that the altered coronary tone may be deleterious in limiting coronary hyperemia in response to anemic hypoxia induced by CO. Meyer and colleagues tried to reproduce the effects of simulated urban CO pollution on the regulation of myocardial perfusion assessed under basal conditions and during the infusion of a β-Adrenergic agonist. They also observed a significant alteration in coronary endothelium-dependent vasorelaxation and an increased uncoupling of eNOS without change in its expression (93).

Cardiac Function in CO Poisoning

Cardiac impairment in CO poisoning has been described mainly in a series of case report studies, however recently Satran and colleagues retrospectively studied in 230 p. treated with hyperbaric oxygen therapy (HBOT) for CO poisoning the cardiac damage (94). In that study Glasgow Coma Scale (GCS) was abnormal in 106 p. (46%), with 50% requiring intubation. Electrocardiogram (ECG) was completely normal in only 16% of p. whereas tachycardia, the most common ECG alteration was present in 41%. Typical ischemic manifestations were present in 30% and nonspecific S-T segment alterations in 41% of p.. The echocardiogram attested an alteration of LV function in 57% of p, with a regional dysfunction prominent in the older group (average age 64 y) with more risk factors as a marker of preexisting coronary disease. A global dysfunction prevailed instead in the younger group (average age 43 y) proving that, in the population with an intact coronary tree, toxic CO doses causes cardiac impairment through the oxidative stress and cellular calcium accumulation secondary to diffuse ischemia with induction of myocardial stunning (95-98). The study proved also that does exist a good concordance between ischemic ECG changes and the markers of myocardial damage (Troponin T and CK-MB). Interestingly the high level of CO-Hb in current smokers lowered the risk of myocardial injury. In a subsequent long-term follow-up (median of 7,6 years) the same cohort of p. showed a long-term increased mortality (24%, a rate three times higher than expected) in p. with positivity for markers of myocardial injury (ECG or TnT) during acute poisoning (99); although the precise mechanism of death is unknown, cardiovascular causes were the most common (44% vs 18%). Kalay et al (100) showed in 20 p. with CO poisoning (mean age 39 years) that the decrease in left ventricular ejection fraction (LVEF) was also negatively correlated with CO-Hb level and CO exposure duration but disappeared more frequently after 24 hours. Another interesting observation was the good correlation between the results of the echocardiogram and TnI determination: all p. with negative echocardiogram were also negative for TnI. A good correlation was also found between brain natriuretic peptide (BNP) and cardiac dysfunction. Since a coronary angiogram was performed in all p. with positive biomarkers showing a normal coronary tree the cardiac dysfunction was attributable only to the ischemic\hypoxic effect of CO. Whereas the description of myocardial infarction is not very uncommon (101,102) in CO poisoning, it was recently described for the first time a case of late stent thrombosis occurring in a 50 years old woman submitted a year before to stent of the left descending coronary artery. She underwent poisoning by accidental exposition to CO during work, presenting with headache, nausea, vomiting followed by sudden syncope (103). COHB was 21,2%, ECG showed 3 mm S-T elevation, TnI was 6.23 ng/ml. The coronary angiogram showed occlusion of the stent; subjected to angioplasty and normobaric oxygen therapy (NBOT) the p. recovered with only minor TnI peak (33,19 ng/ml.) and was discharged after 5 days in hemodinamically stable conditions.

On this basis the recommended study of p. with CO poisoning should include an ECG and a determination of TnT or TnI and possibly BNP as cardiac biomarkers; the positivity of these examinations requires to perform an echocardiogram. In the p. discharged from ICU a periodic follow-up is advisable.

CO Arrhythmias

Acute CO poisoning may cause arrhythmias and, in more severe cases or in cardiac patients, also ventricular fibrillation reducing its threshold (104), but also chronic exposure to low levels of CO may increase the incidence of arrhythmias and worsen cardiovascular diseases augmenting hospital admissions (86). Already in '90 Sheps observed that a level of COHb as low as 6% can increase the incidence of arrhythmias incidence during bicycle exercise in patients with coronary artery disease (CAD) (105). CO is an ubiquitous pollutant and long-term exposure to outdoor air pollution caused by vehicles and industrial emissions increases the risk of mortality from cardiovascular disease by 76% increasing the incidence of ischemia, ventricular arrhythmias and heart failure (106). In the already seen study by Andre (89) CO caused also a proarrhythmogenic effect on β-adrenergic stimulation mediated by a moderate hyperadrenergic state with the consequent Ca^{2+} overload. However, other studies in humans and conscious dogs have failed to observe a change in the incidence of ventricular arrhythmias associated with CO exposure although in different clinical and experimental settings (107-109).

CO has been shown to alter repolarization prolonging QT interval and increasing the variability of Q-T dispersion: in 17 p. with a mild-moderate degree of poisoning Q-Tc interval was prolonged compared to the control group and returned to normal after hospital discharge (110), QT dispersion (QTd) and corrected QTd (cQTd) were also increased. None of the p. underwent arrhythmias probably because they were young, without cardiac disease and the degree of poisoning was lower. Two other very recent studies reported similar results showing a direct correlation between COHb levels (over 21%), Q-Tc prolongation and cardiac enzymes (111); moreover Hanci and colleagues observed an increase in both QTc and in P-wave dispersion (Pwd) that may predispose also to atrial arrhythmias (112). Abramochkin and colleagues (113) performed with microelectrodes the intracellular registration of electrical activity in isolated preparations of atrial and ventricular myocardium, the contraction of atrial stripes was also recorded. CO, administered dissolved in physiological solution, caused different effects in working myocardium, reducing the duration of action potential (APD) and causing a dose-dependent reduction of contractile activity whereas the sino-atrial node activity was accelerated. The inhibition of HO-1-2 by administration of zinc protoporphyrin-IX (ZnPP) caused similar but smaller decrease in electric and contractile parameters showing that the dose of CO used gave rise to concentration superior to the physiological ones. The interference with contractile activity and APD may be partly explained by suppression of ICa_L by CO. One study reports that CO may mimic long Q-T syndrome (LQTS) (114) suggesting a careful evaluation of CO poisoning also as a cause of arrhythmias.

TREATMENT OF CO POISONING

A quick removal from the toxic environment is the first measure to take, protecting adequately the rescue team. When the patient has been removed the administration of high-flow normobaric oxygen via a non-rebreather mask must be provided at least until a target COHb level of 5% is reached (115). The oxygen consumption should be minimized and supportive care undertaken. Unconsciousness, asphyxia, apnea and severe cases require immediate intubation and admission to the intensive care unit (ICU) for monitoring and further study. Because of the risk for the fetus pregnant women must be treated with oxygen for longer periods of time than non-pregnant patients.

The patient should be evaluated as soon as possible for adequacy of ventilation; a blood gas analysis may allow an evaluation of the level of COHb, of the presence of metabolic acidosis related to an altered perfusion and of the adequacy of gas exchanges, representing sometimes an indication to tracheal intubation. A neurological evaluation should be performed whit possible indication to brain imaging. A complete evaluation of cardiovascular function should be performed as already seen.

A severe case of CO poisoning treated by Extracorporeal Membrane Oxygenation (ECMO) has recently been reported: the p. showed a stunned myocardium-induced acute pulmonary edema not responsive to mechanical ventilation (116). Due to severe hypoxemia the p. was submitted to veno-arterial ECMO for three days with a dramatic recovery of hypoxia and of neurological symptoms. The AA. claim that such an aggressive strategy is justified when facing a case of poisoning with acute collapse and refractory respiratory failure.

The best way to carry out oxygen therapy for acute carbon monoxide poisoning has not yet been established: whereas there is a certain indication for normobaric oxygenation (NBO), in case through tracheal intubation, the exact role of hyperbaric oxygen therapy (HBOT), although broadly used, is a debated question and there is no wide agreement about indications, number of sessions and atmospheres absolute (ATA) level.

A series of clinical and experimental data showed that HBO:

1) increases arterial oxygen tension promoting CO elimination (from 320 minutes to 80 with 100% oxygen and to 20 minutes with HBOT at 3,0 ATA)
2) increases dissolved O2 from 0,3 mL/dL to 6,3 mL/dL at 3,0 ATA

3) reduces cerebral edema by causing vasoconstriction
4) increases dissociation of CO from cytochromes and Mb
5) may reduce oxidative stress caused by CO and induce protective genes (117)
6) HBOT at 2,8-3,0 ATA at least for 45 minutes may inhibit β–2 integrins-mediated adherence of neutrophils to the injured endothelium of the central nervous system (CNS) (118)
7) Weaver demonstrated that HBOT could reduce by 46% the incidence of delayed neurological sequelae (119).

The most accepted indications for HBOT are the more severe intoxication (COHb>40%), the loss of consciousness, the complicated poisoning, persistent neurologic dysfunction, cardiac impairment and the poisoning in the pregnant.

Despite these data clinical trials comparing NBO and HBOT failed to provide unequivocal elements in favour of HBOT efficacy with some studies showing some benefits (119-121) and others not showing differences or not recommending it at all (122-124)

Annane and colleagues (125) in a recent study compared the effects of HBOT with NBOT in 385 p. poisoned by CO divided in Trial A, non-comatose, and group B of comatose, having as a primary endpoint the proportion of p. with complete recovery at 1 month. The p. of trial A (179 p.) were randomized to either 6 h of NBO, (arm A0, n = 86) or 4 h of NBO plus one HBO session (arm A1, n = 93). Patients with initial coma (trial B, n = 206) were randomized to either 4 h of NBO plus one HBO session (arm B1, n = 101) or 4 h of NBO plus two 2 HBO sessions (arm B2, n =105).

In non comatose p. there was no evidence for a difference in 1-month complete recovery rates with and without HBO. In Trial B complete recovery was significantly less frequent with two than with one HBO session: on the base of these results the trial was stopped prematurely. The AA concluded that in poisoned p. with transitory loss of consciousness one session of HBOT is not superior to NBO whereas in comatose p. two sessions of HBOT did not show any benefit over one session.

Moreover a recent metanalysis (126) observed significant methodologic and statistical heterogeneity among the trials on HBOT efficacy in the treatment of CO poisoning. Only 6 randomised controlled trials involving 1361 participants were eligible; they does not suggest a significant benefit from HBOT (Odds Ratio for neurological deficits 0.78, 95%CI). The AA conclude that existing randomised trials do not establish whether the administration of HBO to patients with carbon monoxide poisoning reduces the incidence of adverse neurologic outcomes and state that: "Additional research is needed to better define the role, if any, of HBO in the treatment of patients with carbon monoxide poisoning".

One can wonder about so different results on the efficacy of HBOT: we think that the response can be found in the characteristics of HBOT itself that is known to increase oxidative stress, although this latter effect has also positive sides (38,127,128).

Piantadosi (129) studied rats exposed to 10,000 ppm CO for 30 min and subsequently treated with HBOT for 2 h: this treatment at 1.5 ATA reduced ROS level that was, on the contrary, increased by HBOT for 2 h at 2.5 (ATA). The increased H_2O_2 production by monoaminooxidase (MAO) at high oxygen concentration has been identified as the source of excess ROS in this setting. Pun (25) think that it may exist a threshold below which the antioxidant effect of HBOT prevails whereas above it the antioxidant defences are overwhelmed. Naturally this threshold is different in different subjects depending on the antioxidant status, the genetic constitution and the underlying pathological states. If so the problem future studies must solve is not if HBOT therapy is useful but the exact level beyond which HBOT may increase instead than reduce oxidative stress allowing a personalized treatment.

Conclusion

Important advances about the physiologic role of CO as a signalling molecule and the exploitation of the new knowledge in the therapeutic field have been made in the recent years as well as in the toxicology and pathophysiology of CO poisoning. Notwithstanding CO poisoning continue to be the most frequent intoxication and, considered the uncertainties about HBOT efficacy, there has been no therapeutic improvement. New therapies and a precise definition of HBOT role in CO poisoning will be the challenges for the future.

References

[1] Hampson NB, Weaver LK. Carbon monoxide poisoning: a new incidence for an old disease. *Undersea Hyperb Med* 2007;34:163-8.

[2] Harper A, Croft-Baker J: Harper A, Croft-Baker J. Harper A, Croft-Baker J: Carbon monoxide poisoning: undetected by both patients and their doctors. *Age Ageing*. 2004 Mar;33(2):105-9.
[3] Donati SY, Gainnier M,.Chibane-Donati O: Intoxication au CO. *EMC* 2005 :36-986-A-10.
[4] Ernst A, Zibrak JD: Carbon monoxide poisoning. N Engl J Med. 1998 Nov 26;339(22):1603-8.
[5] Boczkowski J, Poderoso JJ, Motterlini R. CO-metal interaction: Vital signaling from a lethal gas. *Trends Biochem Sci.* 2006;31:614–21.
[6] Piantadosi CA: Biological chemistry of carbon monoxide. *Antioxid Redox Signal.* 2002 Apr;4(2):259-70.
[7] Coburn RF, Mayers LB: Myoglobin O2 tension determined from measurements of carboxymyoglobin in skeletal muscle. *Am J Physiol,* 1971;220: 66-74.
[8] Weaver LK. Clinical practice: Carbon monoxide poisoning. *N Engl J Med.* 2009;360:1217–25.
[9] Haldane J: The relation of the action of carbonic oxide to oxygen tension. *J. Physiol.* 1895; 18: 201–17.
[10] Goldbaum LR, Orellano T, Degal E. Mechanisms of the toxic action of carbon monoxide. *Ann. Clin. Lab. Sci.* 1976; 6: 372–6.
[11] Bruce EN, Bruce MC: A multicompartment model of carboxy-hemoglobin and carboxymyoglobin responses to inhalation of carbon monoxide. *J Appl Physiol.* 2003 Sep;95(3):1235-47.
[12] Endeward V, Gros G, Jürgens KD: Significance of myoglobin *Cardiovasc Res.* 2010 Jul 1;87(1):22-9.
[13] Jürgens KD, Papadopoulos S, Peters T, Gros G: Myoglobin: Just an Oxygen Store or Also an Oxygen Transporter? *News Physiol Sci.* 2000 Oct;15:269-274.
[14] Brunori M: Nitric oxide, cytochrome-c oxidase and myoglobin. *Trends Biochem Sci* . 2001 Jan;26(1):21-3.
[15] Gros G, Wittenberg BA, Jue T: Myoglobin's old and new clothes: from molecular structure to function in living cells. *J Exp Biol.* 2010 Aug 15;213(Pt 16):2713-25.
[16] Mammen, P, Kanatous, S. B., Yuhanna, I. S., Shaul, P. W., Garry, M. G., Balaban, R. S, Garry DJ: Hypoxia-induced left ventricular dysfunction in myoglobin-deficient mice. Am. J. Physiol. (2003);285,H2132 -H2141.

[17] Wunderlich C, Flögel U, Gödecke A, Heger J, Schrader J: Acute inhibition of myoglobin impairs contractility and energy state of iNOS-overexpressing hearts. *Circ Res.* 2003 Jun 27;92(12):1352-8.

[18] Brunori M, Vallone B: A globin for the brain. *FASEB J.* 2006 Nov;20(13):2192-7

[19] Nienhaus K, Nienhaus GU: Searching for neuroglobin's role in the brain. *IUBMB Life.* 2007 Aug-Sep;59(8-9):490-7.

[20] Alonso JR, Cardellach F, López S, Casademont J, Miró O: Carbon monoxide specifically inhibits cytochrome c oxidase of human mitochondrial respiratory chain. *Pharmacol Toxicol.* 2003 Sep;93(3):142-6.

[21] Iheagwara KN, Thom SR, Deutschman CS, Levy RJ: Myocardial cytochrome oxidase activity is decreased following carbon monoxide exposure. *Biochim Biophys Acta* 2007; 1772:1112-6.

[22] Piantadosi CA. Carbon monoxide, reactive oxygen signaling, and oxidative stress. *Free Radic Biol Med* 45: 562–569, 2008.

[23] D'Amico G, Lam F, Hagen T, Moncada S: Inhibition of cellular respiration by endogenously produced carbon monoxide. *J Cell Sci.* 2006 Jun 1;119(Pt 11):2291-8.

[24] Borutaite V, Matthias A, Harris H, Moncada S, Brown GC: Reversible inhibition *Am J Physiol Heart Circ Physiol.* 2001 Dec;281(6):H2256-60.

[25] Pun PB, Lu J, Kan EM, Moochhala S: Gases in the mitochondria. *Mitochondrion:* 2010 Mar;10(2):83-93.

[26] Cooper CE, Brown GC: The inhibition of mitochondrial cytochrome oxidase by the gases carbon monoxide, nitric oxide, hydrogen cyanide and hydrogen sulfide: chemical mechanism and physiological significance. *J Bioenerg Biomembr.* 2008 Oct;40(5):533-9.

[27] Blackstone E, Morrison M, Roth MB: H2S induces a suspended animation-like state in mice. *Science* 2005 Apr 22;308(5721):518.

[28] Teodoro RO, O'Farrell PH: Nitric oxide-induced suspended animation promotes survival during hypoxia. *EMBO J.* 2003 Feb 3;22(3):580-7.

[29] Nystul TG, Roth MB: Carbon monoxide-induced suspended animation protects against hypoxic damage in Caenorhabditis elegans. *Proc Natl Acad Sci U S A.* 2004 Jun 15;101(24):9133-6.

[30] Aslami H, Juffermans NP: Induction of a hypometabolic state during critical illness - a new concept in the ICU? *Neth J Med.* 2010 May;68(5):190-8.

[31] Singer M, De Santis V, Vitale D, Jeffcoate W: Multiorgan failure is an adaptive, endocrine-mediated, metabolic response to overwhelming systemic inflammation. *Lancet* 2004; 364:545-8.

[32] Levy RJ: Mitochondrial dysfunction, bioenergetic impairment and metabolic down-regulation in sepsis. *Shock* 2007; 28:24-28.

[33] Baumgart K, Radermacher P, Wagner F: Applying gases for microcirculatory and cellular oxygenation in sepsis: effects of nitric oxide, carbon monoxide, and hydrogen sulfide. *Curr Opin Anaesthesiol.* 2009 Apr;22(2):168-76.

[34] Zegdi R, Perrin D, Burdin M, Boiteau R, Tenaillon A. Increased endogenous carbon monoxide production in severe sepsis. *Intensive Care Med.* 2002;28:793–796.

[35] Hoetzel A, Dolinay T, Schmidt R, Choi AM, Ryter SW: Carbon monoxide in sepsis. *Antioxid Redox Signal.* 2007 Nov;9(11):2013-26

[36] Hui Y, Du J, Tang C, Bin G, Jiang H: Changes in arterial hydrogen sulfide (H_2S) content during septic shock and endotoxin shock in rats. *J Infect.* 2003 Aug;47(2):155-60.

[37] Vreman HJ, Wong RJ, Stevenson DK, Smialek JE, Fowler DR, Li L, Vigorito RD, Zielke HR: Concentration of carbon monoxide (CO) in postmortem human tissues: effect of environmental CO exposure. *J Forensic Sci.* 2006 Sep;51(5):1182-90.

[38] Omaye ST. Metabolic modulation of carbon monoxide toxicity. *Toxicology* 2002;180 : 139–150.

[39] Piantadosi CA, Zhang J, Levin ED, Folz RJ, Schmechel DE. Apoptosis and delayed neuronal damage after carbon monoxide poisoning in the rat. *Exp Neurol* 1997;147:103-14.

[40] Cronje FJ, Carraway 12. MS, Freiberger JJ, Suliman HB, Piantadosi CA. Carbon monoxide actuates O(2)-limited heme degradation in the rat brain. *Free Radic Biol Med* 2004;37:1802-12.

[41] Thom SR: Carbon monoxide-mediated brain lipid peroxidation in the rat. *J Appl Physiol* 1990;68:997-1003.

[42] Watanabe S, Matsuo H, Kobayashi Y, Satoh Y, Fujita M, Nakayama A, Aizawa Y, Shinomiya N, Suzuki S: Transient degradation of myelin basic protein in the rat hippocampus following acute carbon monoxide poisoning. *Neurosci Res.* 2010 Nov;68(3):232-40.

[43] Thom SR, Bhopale VM, Han ST, Clark JM, Hardy KR: Intravascular neutrophil activation due to carbon monoxide poisoning. *Am J Respir Crit Care Med* 2006; 174(11):1239-1248.

[44] Truss NJ, Warner TD: Gasotransmitters and platelets. *Pharmacol Ther.* 2011 Jul 7. [Epub ahead of print].

[45] Peterson SJ, Frishman WH, Abraham NG: Targeting heme oxygenase: therapeutic implications for diseases of the cardiovascular system. *Cardiol Rev.* 2009 May-Jun;17(3):99-111.

[46] Kim HP, Pae HO, Back SH, Chung SW, Woo JM, Son Y, Chung HT: Heme oxygenase-1 comes back to endoplasmic reticulum. *Biochem Biophys Res Commun.* 2011 Jan 7;404(1):1-5.

[47] Yachie A, Niida Y, Wada T, Igarashi N, Kaneda H, Toma T, Ohta K, Kasahara Y, and Koizumi S: Oxidative stress causes enhanced endothelial cell injury in human heme oxygenase-1dcficiency. *J Clin Invest* 1999;103:129–135.

[48] Piantadosi CA: Carbon monoxide poisoning. *Undersea Hyperb Med.* 2004 Spring;31(1):167-77

[49] Ryter SW, Alam J, Choi AM: Heme oxygenase-1/carbon monoxide: from basic science to therapeutic applications. *Physiol Rev.* 2006 Apr;86(2):583-650.

[50] Bauer M, Huse K, Settmacher U, Claus RA: The heme oxygenase-carbon monoxide system: regulation and role in stress response and organ failure. *Intensive Care Med.* 2008 Apr;34(4):640-8.

[51] Abraham NG, Kappas A: Pharmacological and clinical aspects of heme oxygenase. *Pharmacol Rev.* 2008 Mar;60(1):79-127.

[52] Kondo-Nakamura M, Shintani-Ishida K, Uemura K, Yoshida K: Brief exposure to carbon monoxide preconditions cardiomyogenic cells against apoptosis in ischemia-reperfusion. *Biochem Biophys Res Commun.* 2010 Mar 12;393(3):449-54.

[53] Fujimoto H, Ohno M, Ayabe S, Kobayashi H, Ishizaka N, Kimura H, Yoshida K, Nagai R: Carbon monoxide protects against cardiac ischemia–reperfusion injury in vivo via MAPK and Akt–eNOS pathways. *Arterioscler Thromb Vasc Biol* 2004;24:1848-1853.

[54] Wang G, Hamid T, Keith RJ, et al. Cardioprotective and antiapoptotic effects of heme oxygenase-1 in the failing heart. *Circulation.* 2010;121(17):1912–1925.

[55] Piantadosi CA, Carraway MS, Suliman HB: Carbon monoxide, oxidative stress, and mitochondrial permeability pore transition. *Free Radic Biol Med.* 2006 Apr 15;40(8):1332-9.

[56] Chin BY, Jiang G, Wegiel B, Wang HJ, Macdonald T, Zhang XC, Gallo D, Cszimadia E, Bach FH, Lee PJ, Otterbein LE: Hypoxia-inducible factor 1alpha stabilization by carbon monoxide results in cytoprotective

preconditioning. *Proc Natl Acad Sci U S A*. 2007 Mar 20;104(12): 5109-14.

[57] Czibik G, Sagave J, Martinov V, Ishaq B, Sohl M, Sefland I, Carlsen H, Farnebo F, Blomhoff R, Valen G: Cardioprotection by hypoxia-inducible factor 1 alpha transfection in skeletal muscle is dependent on haem oxygenase activity in mice. *Cardiovasc Res*. 2009 Apr 1;82(1):107-14.

[58] Cadenas S, Aragonés J, Landázuri MO: Mitochondrial reprogramming through cardiac oxygen sensors in ischaemic heart disease. *Cardiovasc Res*. 2010 Nov 1;88(2):219-28.

[59] Suliman HB, Carraway MS, Tatro LG, Piantadosi CA: A new activating role for CO in cardiac mitochondrial biogenesis. *J Cell Sci*. 2007 Jan 15;120(Pt 2):299-308.

[60] Lancel S, Hassoun SM, Favory R, Decoster B, Motterlini R, Neviere R: Carbon monoxide rescues mice from lethal sepsis by supporting mitochondrial energetic metabolism and activating mitochondrial biogenesis. *J Pharmacol Exp Ther*. 2009 May;329(2):641-8.

[61] Uji Y, Yamamoto H, Tsuchihashi H, Maeda K, Funahashi T, Shimomura I, et al: Adiponectin deficiency is associated with severe polymicrobial sepsis, high inflammatory cytokine levels, and high mortality. *Surgery* 2009;145:550-7.

[62] Siracusano L, Girasole V: Carbon monoxide and adiponectin in sepsis. *Surgery*. 2010 May;147(5):755.

[63] Wang R. Resurgence of carbon monoxide: an endogenous gaseous vasorelaxing factor. *Can J Physiol Pharmacol* 1998;76:1–15.

[64] Leffler CW, Parfenova H, JH Jaggar: Carbon monoxide as an endogenous vascular modulator. *Am. J. Physiol. Heart Circ. Physiol.* July 1, 2011 301:H1-H11.

[65] Thorup C, Jones CL, Gross SS, Moore LC, Goligorsky S. Carbon monoxide induces vasodilation and nitric oxide release but suppresses endothelial NOS. *Am J Physiol*. 1999;277:F882–F889.

[66] Otterbein, LE et al: Carbon monoxide has anti-inflammatory effects involving the mitogen-activated protein kinase pathway. *Nat. Med.* 6:422–428; 2000.

[67] Bilban M, Haschemi A, Wegiel B, Chin BY, Wagner O, Otterbein LE. Heme oxygenase and carbon monoxide initiate homeostatic signaling. *J Mol Med*. 2008;86:267–279.

[68] Mishra S, Fujita T, Lama VN, Nam D, Liao H, Okada M, Minamoto K, Yoshikawa Y, Harada H, Pinsky DJ: Carbon monoxide rescues ischemic lungs by interrupting MAPK-driven expression of early growth response 1 gene and its downstream target genes. *Proc Natl Acad Sci U S A*. 2006 Mar 28;103(13):5191-6.

[69] Peers C: Ion channels as target effectors for carbon monoxide. *Exp Physiol.* 2011 May 6. [Epub ahead of print].

[70] Wilkinson WJ, Kemp PJ: Carbon monoxide: an emerging regulator of ion channels. *J Physiol.* 2011 Jul 1;589(Pt 13):3055-62.

[71] Peers C, Steele DS: Carbon monoxide: A vital signalling molecule and potent toxin in the myocardium. *J Mol Cell Cardiol.* 2011 May 26. [Epub ahead of print].

[72] Foresti R, Bani-Hani MG, Motterlini R: Use of carbon monoxide as a therapeutic agent: promises and challenges. *Intensive Care Med.* 2008 Apr;34(4):649-58.

[73] Motterlini R, Otterbein LE: The therapeutic potential of carbon monoxide. *Nat Rev Drug Discov*. 2010 Sep;9(9):728-43.

[74] Ryter SW, Choi AM: Heme oxygenase-1/carbon monoxide: from metabolism to molecular therapy. *Am J Respir Cell Mol Biol.* 2009 Sep;41(3):251-60.

[75] Amitai Y, Zlotogorski Z, Golan-Katzav V, Wexler A, Gross D: Neuropsychological impairment from acute low-level exposure to carbon monoxide. *Arch Neurol* 1998;55:845–848.

[76] Hurley RA, Hopkins RO, Bigler ED, Taber KH: Applications of functional imaging to carbon monoxide poisoning. *J Neuropsychiatry Clin Neurosci*. 2001 Spring;13(2):157-60.

[77] Lo CP, Chen SY, Lee KW, Chen WL, Chen CY, Hsueh CJ, Huang GS: Brain injury *AJR Am J Roentgenol*. 2007 Oct;189(4):W205-11.

[78] Gandini C, Castoldi AF, Candura SM, Locatelli C, Butera R, Priori S, Manzo L: Carbon monoxide cardiotoxicity. *J Toxicol Clin Toxicol.* 2001;39(1):35-44.

[79] Henry TD, Lesser JR, Satran D. Myocardial fibrosis from severe carbon monoxide poisoning detected by cardiac magnetic resonance imaging. *Circulation* 2008;118:792.

[80] Doherty S: History, pathophysiology, clinical presentation and role of HBO in acute CO poisoning .*Emergency Medicine*: 2000:12; 55.

[81] Goldstein M: Carbon monoxide poisoning. *J Emerg Nurs*. 2008 Dec;34(6):538-42.

[82] Gabrielli A, Layon AJ: Carbon Monoxide Intoxication during Pregnancy: A Case Presentation and Pathophysiologic Discussion, with Emphasis on Molecular Mechanisms. *J Clin Anesth.* 1995 Feb;7(1):82-7.

[83] Fukuto JM, Collins MD: Interactive endogenous small molecule (gaseous) signaling: implications for teratogenesis. *Curr Pharm Des.* 2007;13(29):2952-78.

[84] Kao LW, Nañagas KA: Toxicity associated with carbon monoxide. *Clin Lab Med.* 2006 Mar;26(1):99-125.

[85] Hampson NB, Hauff NM: Carboxyhemoglobin levels in carbon monoxide poisoning: do they correlate with the clinical picture? *Am J Emerg Med.* 2008 Jul;26(6):665-9.

[86] Adams KF, Koch G, Chatterjee B, Goldstein GM, O'Neil JJ, Bromberg PA, et al. Acute elevation of blood carboxyhemoglobin to 6% impairs exercise performance and aggravates symptoms in patients with ischemic heart disease. *J Am Coll Cardiol* 1988;12:900–9.

[87] Penney DG: Hemodynamic response to carbon monoxide. *Environ Health Perspect.* 1988 Apr;77:121-30.

[88] Suner S, Jay G: Carbon monoxide has direct toxicity on the myocardium distinct from effects of hypoxia in an ex vivo rat heart model. *Acad Emerg Med.* 2008 Jan;15(1):59-65.

[89] Andre L, Boissiere J, Reboul C, Perrier P, Zalvidea S, Thireau J, Meyer G, Bideaux P, Obert P, Cazorla O, Richard S: Carbon monoxide pollution promotes cardiac remodeling and ventricular arrythmia in healthy rats. *Am J Respir Crit Care Med* 2010;181. 587–595, 181:587–595.

[90] Li Volti G, Vanella L, Gazzolo D, Galvano F: Carbon monoxide: vasoconstrictor or vasodilator? That's the question. *Am J Physiol Renal Physiol.* 2008 Oct ; 295 (4):F901-3.

[91] Lamon BD, Zhang FF, Puri N, Brodsky SV, Goligorsky MS, Nasjletti A: Dual pathways of carbon monoxide-mediated vasoregulation: modulation by redox mechanisms. *Circ Res.* 2009 Oct 9;105(8):775-83.

[92] Favory R, Lancel S, Tissier S, Mathieu D, Decoster B, Nevière R: Myocardial dysfunction and potential cardiac hypoxia in rats induced by carbon monoxide inhalation. *Am J Respir Crit Care Med.* 2006 Aug 1;174(3):320-5.

[93] Meyer G, Boissiere J, Tanguy S, Rugale C, Gayrard S, Jover B, Obert P, Reboul C: Carbon Monoxide Pollution Impairs Myocardial Perfusion Reserve: Implication of Coronary Endothelial Dysfunction. COHb level

in smokers is a risk factor for ischemic heart disease. *Cardiovasc Toxicol.* 2011 Jul 27.

[94] Satran D, Henry CR, Adkinson C, Nicholson CI, Bracha Y, Henry TD. Cardiovascular manifestations of moderate to severe carbon monoxide poisoning. *J Am Coll Cardiol.* 2005; 45: 1513–1516.

[95] Tritapepe L, Macchiarelli G, Rocco M, Scopinaro F, Schillaci O, Martuscelli E, Motta PM: Functional and ultrastructural evidence of myocardial stunning after acute carbon monoxide poisoning. *Crit Care Med.* 1998 Apr;26(4):797-801.

[96] Fotbolcu H, Incedere O, Bakal RB, Tanalp AC, Astarcioglu MA, Dindar I: Reversible myocardial stunning due to carbon monoxide exposure. *Cardiovasc J Afr.* 2011 Mar-Apr;22(2):93-5.

[97] Yanir Y, Shupak A, Abramovich A, Reisner SA, Lorber A: Cardiogenic shock complicating acute carbon monoxide poisoning despite neurologic and metabolic recovery. *Ann Emerg Med.* 2002 Oct;40(4):420-4.

[98] Rastelli G, Callegari S, Locatelli C, Vezzani G: Myocardial injury in carbon monoxide poisoning. *G Ital Cardiol* (Rome). 2009 Apr;10(4):227-33.

[99] Henry CR, Satran D, Lindgren B, Adkinson C, Nicholson CI, Henry TD: Myocardial injury and long-term mortality following moderate to severe carbon monoxide poisoning. *JAMA* 2006; 295:398–402.

[100] Kalay N, Ozdogru I, Cetinkaya Y, Eryol NK, Dogan A, Gul I, Inanc T, Ikizceli I, Oguzhan A, Abaci A: Cardiovascular effects of carbon monoxide poisoning. *Am J Cardiol.* 2007 Feb 1;99(3):322-4.

[101] Grieb G, Simons D, Piatkowski A, Altiok E, Eppstein RJ, Bernhagen J, Pallua N: Carbon monoxide intoxication versus myocardial infarction: An easy diagnosis? *Burns.* 2011 Jun;37(4):e29-31.

[102] Tucciarone M, Dileo PA, Castro ER, Guerrero M: Myocardial infarction *Am J Ther.* 2009 Sep-Oct;16(5):462-5.

[103] Dileo PA, Tucciarone M, Castro ER, Guerrero M: Late stent thrombosis secondary to carbon monoxide poisoning. *Cardiovasc Revasc Med.* 2011 Jan-Feb;12(1):56-8.

[104] Aronow WS, Stemmer EA, Wood B, Zweig S, Tsao KP, Raggio L: Carbon monoxide and ventricular fibrillation threshold in dogs with acute myocardial injury. *Am Heart J.* 1978 Jun;95(6):754-6.

[105] Sheps DS, Herbst MC, Hinderliter AL, Adams KF, Ekelund LG, O'Neil JJ, Goldstein GM, Bromberg PA, Dalton JL, Ballenger MN, Davis SM, Koch GG: Production of arrhythmias by elevated carboxyhemoglobin

in patients with coronary artery disease. *Ann. Intern. Med.* 1990; 113, 343–351.

[106] Miller KA, Siscovick DS, Sheppard L, Shepherd K, Sullivan JH, Anderson GL, Kaufman JD: Long-term exposure to air pollution and incidence of cardiovascular events in women. *N Engl J Med* 2007; 356:447–458.

[107] Kizakevich PN, McCartney ML, Hazucha MJ, Sleet LH, Jochem WJ, Hackney AC, Bolick K: Noninvasive ambulatory assessment of cardiac function in healthy men exposed to carbon monoxide during upper and lower body exercise. *Eur J Appl Physiol.* 2000 Sep;83(1):7-16.

[108] Dahms TE, Younis LT, Wiens RD, Zarnegar S, Byers SL, Chaitman BR: Effects of carbon monoxide exposure in patients with documented cardiac arrhythmias. *J Am Coll Cardiol.* 1993 Feb;21(2):442-50.

[109] Vanoli E, De Ferrari GM, Stramba-Badiale M, Farber JP, Schwartz PJ: Carbon monoxide and lethal arrhythmias in conscious dogs with a healed myocardial infarction. *Am Heart J.* 1989 Feb;117(2):348-57.

[110] 110) Gürkan Y, Canatay H, Toprak A, Oral E, Toker K. Carbon monoxide poisoning- a cause of increased QT dispersion. *Acta Anesthesiol Scand* 2002; 46: 180-3.

[111] Yelken B, Tanriverdi B, Cetinbaş F, Memiş D, Süt N: The assessment of QT intervals in acute carbon monoxide poisoning. *Anadolu Kardiyol Derg.* 2009 Oct;9(5):397-400.

[112] Hanci V, Ayoğlu H, Yurtlu S, Yıldırım N, Okyay D, Erdoğan G, Abduşoğlu M, Yılmaz M, Ozkoçak I: Effects of acute carbon monoxide poisoning on the P-wave and QT interval dispersions. *Anadolu Kardiyol Derg.* 2011 Feb;11(1):48-52.

[113] Abramochkin DV, Haertdinov NN, Porokhnya MV, Zefirov AL, Sitdikova GF: Carbon monoxide affects electrical and contractile activity of rat myocardium. *J Biomed Sci.* 2011 Jun 15;18:40.

[114] Onvlee-Dekker IM, De Vries AC, Ten Harkel AD: Carbon monoxide poisoning mimicking long-QT induced syncope. *Arch Dis Child* 2007;92:244–5.

[115] Ilano AL, Raffin TA: Management of carbon *Chest.* 1990 Jan; 97(1):165-9.

[116] Wang YT, Chen CW, Chian CF, Perng WC, Hong GJ, Su WL: Extracorporeal Membrane Oxygenation for Management of Carbon Monoxide Intoxication *JMed Sci* 2010;30(3):101-105.

[117] Godman CA, Joshi R, Giardina C, Perdrizet G, Hightower LE: Hyperbaric oxygen treatment induces antioxidant gene expression. *Ann N Y Acad Sci.* 2010 Jun;1197:178-83.

[118] Thom SR, Mendiguren I, Hardy K, Bolotin T, Fisher D, Nebolon M, Kilpatrick L: Inhibition of human neutrophil □2-integrin dependent adherence by hyperbaric O2. *Am J Physiol* 1997;272:C770–C777.

[119] Weaver LK, Hopkins RO, Chan KJ, et al. Hyperbaric oxygen for acute carbon monoxide poisoning. *N Engl J Med* 2002; 347: 1057–1067.

[120] Ducassé JL, Celsis P, Marc-Vergnes JP: Non-comatose patients with acute carbon monoxide poisoning: hyperbaric or normobaric oxygenation? *Undersea Hyperb Med.* 1995 Mar;22(1):9-15.

[121] Thom SR, Taber RL, Mendiguern II, Clark JM, Hardy KR, Fisher AB: Delayed neuropsychologic sequelae after carbon monoxide poisoning: prevention by treatment with hyperbaric oxygen. *Ann Emerg Med* 1995;25:474–480.

[122] Raphael JC, Elkharrat D, Jars-Guincestre MC, Chastang C, Chasles V, Vercken JB, Gajdos P: Trial of normobaric and hyperbaric oxygen for acute carbon monoxide intoxication. *Lancet.* 1989 Aug 19; 2(8660):414-9.

[123] Mathieu D, Mathieu-Nolf M, Wattel F (1996) Intoxications par le monoxide de carbone: aspects actuels. *Bull Acad Natl Med* 180:965–971.

[124] Scheinkestel CD, Bailey M, Myles PS,Jones K, Cooper DJ, Millar IL, Tuxen DV: Hyperbaric or normobaric oxygen for acute carbon monoxide poisoning: a randomised controlled clinical trial. *Med J Aust* (1999) ;170:203–210.

[125] Annane D, Chadda K, Gajdos P, Jars-Guincestre MC, Chevret S, Raphael JC: Hyperbaric oxygen therapy for acute domestic carbon monoxide poisoning: two randomized controlled trials. *Intensive Care Med.* 2011 Mar;37(3):486-92.

[126] Buckley NA, Juurlink DN, Isbister G, Bennett MH, Lavonas EJ: Hyperbaric oxygen for carbon monoxide poisoning. *Cochrane Database Syst Rev.* 2011 Apr 13;(4):CD002041.

[127] Mannaioni PF, Vannacci A, Masini E:Carbon monoxide: the bad and the good side of the coin, from neuronal death to anti-inflammatory activity. *Inflamm Res contro HBOT ros* 2006;55:261-73.

[128] Thom SR: Oxidative stress *J Appl Physiol.* 2009 Mar;106(3):988-95

[129] Piantadosi, C.A., Tatro, L., Zhang, J: 1995. Hydroxyl radical production in the brain after CO hypoxia in rats. *Free Radical Biol. Med.* 18, 603–609.

In: Carbon Monoxide: Sources, Uses and Hazards ISBN:978-1-61942-055-7
Editors: D. DiLoreto et al.

Chapter 2

CARBON MONOXIDE: SOURCES, USES AND HAZARDS

Hueng-Chuen Fan[1], Chun-Jung Juan[2] and Shyi-Jou Chen[1*]

[1]Department of Pediatrics Tri-Service General Hospital, National Defense Medical Center, Taipei City,Taiwan

[2]Department of Radiology, Tri-Service General Hospital, National Defense Medical Center, Taipei City,Taiwan

ABSTRACT

Carbon monoxide (CO) intoxication is one of the main causes of mortality in the world because of its tastelessness, colorlessness, and odorlessness without irritating properties, leading to unawareness of excessive inhalation of potentially lethal concentrations for victims. Moreover, the frequency of non-lethal CO intoxications is even higher but difficult to measure for the nonspecific presentations caused by this intoxicant. In addition, CO measurements at the time of actually reaching the emergency unit may not always correlate with symptoms, findings, and prognosis in acute CO intoxications. Although non-invasive and quick brain imaging, including computerized tomography (CT) and magnetic resonance imaging (MRI), may help clinicians to approach

[*] Correspondence: S.J. Chen Department of Pediatrics, Tri-Service General Hospital, National Defense Medical Center, 325, Section 2, Cheng-Kung Road, Neihu, Taipei 114, Taiwan. Email address: pedneuchen@hotmail.com. (SJ Chen)

typical patients with CO intoxication, however, they are not always sensitive to patients with varied extent of CO inhalation. Given the universal use of fossil energy sources, all of the individuals are dangerously threatened by this "invisible killer". Understanding the causes, risks, and prevention of CO intoxication may decrease needless mortality and morbidity.

INTRODUCTION

CO is the leading cause of poisoning mortality and accounts for an estimated 50,000 people each year seek medical attention or miss work in the USA [1, 2]. Most accidental deaths related to CO intoxication are due to house fires and automobile exhaust[3]. Non fire-related CO deaths in 2007 were associated with indoor heating systems (70%) [4]. In Taiwan, hospitalizations due to CO poisoning increased from 261 people in 1997 to 740 people in 2003 and most patients with CO poisoning are associated with misuse of the gas heater, which was especially located either in the bathroom or in the closed balcony [5].

Carbon monoxide (CO) is a colorless, odorless, and tasteless environmental toxicant gas [6-8]. Common sources of CO include motor vehicle exhaust, burning of charcoal, wood, kerosene, or natural gas for heating and cooking [9]. CO intoxications can involve multiple organ systems (e.g. lung, peripheral and central nervous system) and undetected or unsuspected CO exposure can be fatal [10]. When a patient with CO poisoning is suspected, clinical presentations and imaging features may be diverse and medical courses of the same intoxication source can differ significantly [11]. As CO intoxication is often misinterpreted by both patients and physicians [12], a better understanding of pathophysiologic mechanisms of CO poisoning may result in therapeutic advances.

PATHOPHYSIOLOGY

The sources of CO can be divided into endogenous and exogenous. Functionally, endogenous CO involves inflammation, apoptosis, cell proliferation, and mitochondrial biogenesis [2, 13]. Exogenous CO is absorbed exclusively by the respiratory tract and diffuses through the alveolo-capillary membrane in way similar to oxygen and increase exposure may cause intoxication. Initially, the pathophysiology of CO poisoning was contributed to depriving oxygen through the formation of a reversible complex between CO and the hemoglobin, carboxyhemoglobin (CO-Hb) [14], because the affinity of CO for heme protein is approximately 240 times that of oxygen. Therefore, the CO-Hb complex is unable to transport as much oxygen as normal Hb, causing tissue hypoxia [15] and only a small concentration of CO in inspired air can easily take the place of oxygen in circulating Hb [16]. Some practitioners may access the severity of CO intoxication according to the level of CO-Hb and may recommend hyperbaric oxygen therapy (HBO) when the level of CO-Hb is greater than 25%[17, 18]. CO-Hb levels in fact should not be a predictive factor of the severity of illness, duration of exposure, and prognosis of patients with CO poisoning [19]. An animal experiment [10] showed that dogs breathing 13% CO died within 1 hour when CO-Hb levels elevated from 54% to 90%. Exchange transfusion with blood containing 80% CO-Hb to otherwise healthy dogs resulted in no toxic effects, but the level of CO-Hb levels increased from 57% to 64%, suggesting effects of CO may activate multiple pathways that are independent of the pathways to hypoxia, such as binding to cytochrome P450 enzymes, myoglobin, guanylyl cyclase, nitric oxide (NO) synthases, and cytochrome *c* oxidase (complex IV), leading to cell necrosis [20] and apoptosis[21], culminating a variety of manifestations in patients with CO poisoning.

Cytochrome P450 superfamily (CYP): The cytochrome P450 enzyme system (CYP) consists of a superfamily of hemoproteins that catalyse the oxidative metabolism of a wide array of exogenous chemicals including drugs, carcinogens, toxins and endogenous compounds such as steroids, fatty acids and prostaglandins[22]. The name cytochrome P450 is derived from the fact that cytochrome P450 has a heme group and it bounds CO to set off a characteristic absorbance maximum at 450 nm [23, 24]. However, the physiologic significance of the binding is not known. An in vivo study shows that ethanol-derived oxidative damage (especially CYP 2E1) on human hepatocytes through heme oxygenase-1 (HO-1) induction. Heme degradation and CO release may mediate the protective effects through inhibiting ethanol-induced CYP synthesis and enzymatic activity [25]. Obviously, CO poisoning may interfere with liver functions via the interaction with CYP.

Myoglobin: Myoglobin is a cytoplasmic hemoprotein, expressed solely in cardiac myocytes and oxidative skeletal muscle fibers. Structurally, the backbone of myoglobin consists of eight α-helices that wrap around a central pocket containing a heme group, which is capable of binding various ligands including oxygen, CO and NO [26]. Functionally, myoglobin is an O_2-storage protein in muscle, capable of releasing O_2 during periods of hypoxia or anoxia. CO binds to cardiac and skeletal muscle's myoglobin, with cardiac muscle taking up about three times as much as skeletal muscle. CO binding to myoglobin may reduce oxygen availability in the heart and lead to arrhythmias and cardiac dysfunction and may contribute to direct skeletal muscle toxicity and rhabdomyolysis [3].

Soluble guanylyl cyclase (sGC): The sGC subunits have three distinct regions: sGCβ1 containing a N-terminal NO sensing heme domain, a central dimerization region, and a C-terminal catalytic guanylyl cyclase domain[27]. sGCα1 has a similar subunit arrangement. Its N-terminal domain does not contain a heme. The heme domain in the sGC can catalyze the conversion of guanosine 5'-triphosphate (GTP) to guanosine 3',5'-cyclic monophosphate (cGMP) [28]. CO and NO both can regulate sGC, because the sGC contains a NO/CO sensing and catalytic complex and the NO–cGMP signaling pathway is involved in cardiovascular and neurological processes and sGC activators can potentially treat hypertension, erectile dysfunction, congestive heart failure, and chronic renal disease [29, 30]. When compared with NO, CO has a much lower affinity and weak stimulatory properties for sGC [31]. A report shows that exogenously added CO inhibits platelet aggregation and relaxes isolated blood vessel preparations by elevating intracellular levels of cGMP [32]. However, the differential regulation of sGC by NO and CO in terms of affinity, activity, and coordination states demonstrates an intriguing complexity of sGC. As the activity of CO-bound sGC is to bind to the iron atom of the heme moiety, thereby activating the enzyme and increasing intracellular cGMP production, it is speculated that loss of consciousness in an animal model of CO poisoning may be associated with this mechanism [33, 34].

Nitric oxide (NO) synthases: The role of NO has been researched extensively in the setting of CO poisoning. NO is a naturally occurring vasodilator, which may result in systemic hypotension [35]. Animal studies showed cerebral vasodilatation after exposure to CO caused loss of consciousness and increased NO levels [36]. Administration of NO synthase inhibitors showed the inhibition of the cerebral vasodilatation [37]. NO may affect the adherence of neutrophils to the endothelium by targeting at an adhesion molecule, the B_2-integrin [38, 39]. Neutrophil adherence to the microvasculature may activate xanthine oxidase, cause oxidative radical formation and oxidative damage, and induce brain lipid peroxidation, which is thought to be responsible for delayed neurologic sequelae (DNS) [38-43]. However, NO is also a signaling molecule[44]. With CO poisoning, the amount of NO in the cell interior rises because CO usurps the spot of NO on the heme proteins, leading to the excessive generation of tissue-damaging oxidants and free radicals[16], which trigger a cascade of events resulting in neurologic and cardiac injury.

Mitochondria: Cytochrome *c* oxidase (complex IV) serves as the mitochondrial enzyme responsible for the reduction of oxygen into water as the final step of the electron transport chain. When CO binds to cytochrome c oxidase, it causes mitochondrial dysfunction and cellular respiratory impairment, leading to generation of oxygen radicals, NO, and mitochondrial-derived ROS [10, 45], which linked to the activation of various adaptive signaling pathways, including the mitogen-activated protein kinase family members, JNK and p38, and NF-κB [46-48]. All these relevant cascades are activated and triggered and result in oxidative stress related damage [15].

Clinical Manifestations

Acute Effects

High oxygen-demand organs, such as brain and heart, are most sensitive to the toxic effects of CO[49]. Early neurologic manifestations include dizziness, headache. When the CO concentration increases, patients may demonstrate more severe symptoms, such as altered mental status, confusion, syncope, seizure, acute strokelike syndromes, and coma. Acute CO poisoning may cause lesions in various brain regions, including cerebral cortex, the globus pallidus, the hippocampus, the basal ganglia and the caudate putamen [50]. Early cardiovascular effects of CO poisoning manifest as a response to hypoxia [51], resulting in electrocardiographic QT prolongation, myocardial ischemia, ventricular arrhythmia or atrial thrombus formation [52, 53]. More significant exposures result in hypotension, dysrhythmia, ischemia, infarction, and, in extreme cases, cardiac arrest. Early deaths after CO exposure may be due to cardiac dysrhythmias [54]. Myocardial infarction can occur in CO poisoning in the absence of underlying coronary disease [55]. Hypotension may result from myocardial injury secondary to hypoxia-ischemia, direct myocardial depressant activity from myoglobin binding, peripheral vasodilatation, or a combination [2]. The presence of systemic hypotension in CO poisoning may be related to the severity of CNS structural damage [56-60]. Skin and mucous membrane is also a target of CO poisoning, including blisters [61] and cherry-red mucous membranes that may appear when CO-Hb levels are between 50 and 60%, and lips, skin and mucous membranes are more common sites in post-mortem [16]. CO poisoning also may result in rhabdomyolysis and acute renal failure [62-64] and non-cardiogenic pulmonary edema [65, 66]. CO intoxication can mimic symptoms of acute myocardial infarctions to make the differential diagnosis difficult and complex [67]. The acute effects produced by CO related to ambient concentration in parts per million (ppm) are listed in Table 1.

Table 1. Concentration of CO-Hb and clinical presentations

Conc. (ppm)	Symptoms
35 ppm	Headache and dizziness within 6 to 8 hours of constant exposure.
100 ppm	Slight headache in 2 to 3 hours.
200 ppm	Slight headache within 2 to 3 hours; loss of judgment.
400 ppm	Frontal headache within 1 to 2 hours.
800 ppm	Dizziness, nausea, and convulsions within 45 minutes; insensible within 2 hours.
1,600 ppm	Headache, tachycardia, dizziness, and nausea within 20 minutes; death in less than 2 hours.
32,00 ppm	Headache, dizziness and nausea in 5 to10 minutes. Death within 30 minutes.
64,00 ppm	Headache and dizziness in 1 to 2 minutes; convulsions, respiratory arrest, and death in less than 20 minutes.
128,00 ppm	Unconsciousness after 2 to 3 breaths. Death in less than 3 minutes.

Adapted from [170].

Delayed Effects

The effects of CO are not confined to the period immediately after exposure. Animal studies showed that CO poisoning causes transient degradation of myelin basic protein (MBP) and axonal injury in the hippocampus [68] reacted with malonyaldehyde, a product of lipid peroxidation, to trigger immunologic cascades, leading to delayed CO-mediated neuropathology [69]. Symptoms related to DNS are varied, including dementia, amnesia, psychosis, chorea, memory loss, confusion, ataxia, seizures, urinary and fecal incontinence, emotional labiality, disorientation, hallucinations, parkinsonism, mutism, cortical blindness, psychosis, and gait and other motor disturbances [70-74]. In general, patients who present with more symptoms at acute phase are the most likely to develop DNS. The incidence of DNS, caused by CO intoxication ranges from 2.75% to 40% of patients after CO poisoning [72, 74-77]. Risk factors for developing DNS include age (>36 years), longer duration of exposure to CO, higher of carboxyhemoglobin (CO-Hb) (>10% or 25%) and damage in the globus pallidus or white matter [78-81]. The lucid interval between recovery from the initial exposure and the development of DNS was 2 to 40 days (mean 22.4 days). The large variability in incidence may be due to the definitions and neuropsychometric tests by different investigators. EEG changes and coma, with an incidence of 32–67% [82]. The findings of neuroimaging in subacute CO poisoning patients are variable. However, neuropsychiatric symptoms after CO poisoning may possibly contribute to demyelination of cerebral white matter fibers [83]. New MR imaging techniques, such as DWI, can demonstrate the damage in the brain in patients with CO poisoning [83, 84], such as bilateral periventricular white matter, the splenium of the corpus callosum, internal capsules, and brainstem showing moderately decreased ADC values. Consistently, single-photon emission CT scans (SPECT) also reveal increased hypoperfusion in bilateral white matter and some parts of the cerebral cortex, which correlated well with the DWI findings [85]. Positron emission tomography (PET) may show decreased glucose metabolism in the basal ganglia [86]. SPECT showed a better correlation with neurologic sequlae, such as development of delay when compared with other neuroimaging findings [85, 87].

Chronic Effects

Patients with CO poisoning should be regularly followed up after discharge. The extent of sequelae after CO exposure are unpredictable and can be permanent. Statistical data showed that approximately 6 years after poisoning, 19% of patients had cognitive problems [88] and 37% had abnormal neurologic evaluations [89]. Prospective studies of patients treated with normobaric oxygen (NBO) showed that 34% reported symptoms such as headaches or memory problems at 4 weeks [76] and 46% had neuropsychological sequelae at 6 weeks [18]. Symptoms of chronic poisoning may differ from those of acute poisoning, including chronic fatigue, affective conditions and emotional distress, memory deficits, difficulty working, sleep disturbances, vertigo, neuropathy, paresthesias, recurrent infections, polycythemia, abdominal pain, and diarrhea as headache, dizziness, anorexia, apathy, insomnia, personality disturbance, and cognitive impairment [90]. Warehouse workers chronically exposed to CO from propane combustion may have intermittent headache, nausea, and light-headedness [19, 91]. Highway toll workers have trouble performing parallel processing tasks with documented continued exposure to low levels of CO [92]. An unusual case report noted prominent verbal memory, visual recall, and learning problems after three years exposure to 180 ppm CO from a faulty furnace [93]. In addition, chronic CO exposure can exacerbate the preexisting diseases like ischemic heart disease or dementia [12], accelerate atherosclerosis [94], and associate polycythemia and cardiomegaly, likely secondary to chronic hypoxia [51]. Other sequelae include gait and motor disturbances, peripheral neuropathy, hearing loss and vestibular abnormalities, and dementia and psychosis, which can be permanent [13]. Although symptoms may spontaneously resolve in most patients [95], clinicians and nurses should be vigilant for patients with suspicious chronic CO exposure to prevent from continued catastrophic outcomes.

Special Populations

Infants and children may be more susceptible to CO poisoning than adults because of their rapid respiratory rate, higher metabolic rate, and oxygen uptake [96, 97], Moreover, CO binds much more tightly to fetal than adult hemoglobin so that occult CO poisoning may present as an acute life-threatening in infants [98, 99]. This may explain why pregnant patients with only moderate symptoms and no syncope have had devastating fetal outcomes. Symptoms in pediatric patients are often nonspecific, such as nausea and vomiting, and can be mistakenly diagnosed as a viral illness [98]. An increased incidence of syncope and lethargy is reported in the pediatric patients when compared with adults [97]. CO can cross the placenta, and earlier gestational age of the fetus during CO exposure has been associated with anatomic malformations, whereas poor neurologic development and functional disturbances are reported after CO exposure at any gestational age [100-102]. Fetal mortality exceeds 50% in cases of severe poisoning [103].Therefore, when maternal exposure is more severe, fetal complications, such as stillbirth, anatomic malformations, and neurologic disability, more often occur [101, 103-105]. Autopsy of fetal demise of CO poisoning female rhesus monkey showed that fetal brains were markedly swollen and extensive hemorrhagic necrosis, particularly in the basal ganglia and globus pallidus [106], which are compatible with neuroimaing findings in patients with CO intoxication (see later).

DIAGNOSIS

History

CO poisoning is usually diagnosed on the basis of a suggestive history. However, the correct diagnosis of CO poisoning is not easy. In addition to the history from patients, collateral history from attending paramedics or cohabitants with or without similar symptoms, apparent in certain circumstances, such as the smoke-inhalation victim or the unconscious patient in a garage with a running automobile may give clues to physicians. In a situation that a patient presents unconsciousness or other conditions and cannot give a complete history of exposure, an astute physical examination, detailed laboratory tests and brain imagings, and experienced stuff are the key to a corrective and early diagnosis of CO poisoning. CO intoxication should be included in the differential diagnosis of all patients presenting to any physician with headache, dyspnea, nausea or altered level of consciousness, thus making treatment appropriate and timely.

Carboxyhemoglobin (CO-Hb) and Other Tests

In general, patients with CO poisoning may present metabolic acidosis, presumably from a combination of hypoxia, inhibition of cellular respiration, and increased metabolic demand. However, clinical manifestations are unpredictable. A patient may not show any clinical toxicity with a high level of CO-Hb after a short exposure of CO[107], whereas a patient with similar CO-Hb level after a prolonged lower level exposure manifest serious symptoms. The severity of clinical symptoms is related not only to the concentration of CO, but also to the duration of exposure [108, 109]. Importantly, CO-Hb levels decline with time and with oxygen therapy, and therefore an initial CO-Hb level may not reflect accurately the magnitude of a patient's exposure if the blood is drawn a period of time after CO exposure or after oxygen therapy. Even so, whenever CO intoxication is suspected at emergent room, CO-Hb levels should be measured, It is known that low CO-Hb levels (<15–20%) correlate with mild symptoms, such as nausea and headache. Levels greater than 60% to 70% are usually rapidly fatal. Intermediate levels often are not correlated significantly with symptoms or with prognosis. Also, it needs to address that a nonsmoker may have a baseline level of less than 1% to 3% from endogenous production and background environmental exposure, whereas smokers may have levels of 10%, slightly higher immediately after smoking [110, 111]. Arterial or venous sampling is no difference because a prospective comparison of arterial and venous CO-Hb levels in poisoned patients has shown similar [112]. CO-Hb levels should be measured via a co-oximeter, which measures total hemoglobin concentration, oxyhemoglobin, and deoxyhemoglobin and concentrations of abnormal hemoglobins, such as CO-Hb and methemoglobin, by differentiating wavelength absorbance values [113], because routine blood gas analyzers without co-oximeters cannot recognize the contribution of abnormal hemoglobins and pulse oximetry is unreliable in estimating O_2Hb saturation in CO-exposed patients [114-116]. Therefore, data from pulse oximetry and should be interpreted with caution. Clinical manifestations of CO poisoning in relation with levels of CO-Hb are listed in Table 2.

Table 2. Clinical picture of CO poisoning in relation with levels of carboxyhaemoglobin

CO-Hb (%)	Symptoms
0-10	Asymptomatic
10-20	Weakness, headache
20-30	Severe headaches, nausea, vertigo
30-40	Nausea, vomiting, blurred vision, muscle weakness
40-50	Loss of consciousness, tachypnoea, tachycardia
50-60	Coma, convulsions
>60	Cardiovascular collapse, respiratory distress

Adapted from [162, 171].

Moreover, the level of acidosis is correlated with exposure duration, severity of clinical symptoms, or adverse sequelae after CO poisoning [117]. Lactate has been a marker for severe poisoning [118]. Chest radiograph may show evidence of non-cardiogenic pulmonary edema. EKG may demonstrate nonspecific changes, dysrhythmias, or changes associated with myocardial ischemia. Cardiac markers and creatine phosphokinase may be elevated. Fetal monitoring and other tests of fetal well-being may detect fetal compromise in a CO-poisoned pregnant patient.

Imaging

Computed tomography (CT) of the brain in patients with severe CO exposure may show single or bilateral, round, low-density areas, especially in the inner parts of the globus pallidus [78, 119]. Development of these lesions has been correlated with local decreasing blood flow, metabolic acidosis, and hypotension during CO poisoning in animal models [56-58]. However, these lesions are not specific for CO poisoning and may be found in other intoxications, such as methanol [120]or hydrogen sulfide [121] poisoning. Therefore, in addition to CO intoxication, clinician should consider other causes. Globus pallidus lesions may be delayed several days after initial presentation [122] and may resolve spontaneously [123]. Concomitant white matter lesions [83] and cerebellum [124] (Figure 1) also may be seen.

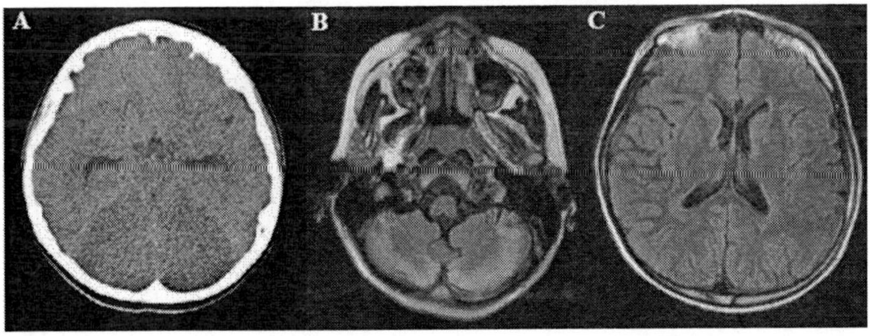

Figure 1. Axial non-enhanced cranial CT scan on the day of CO poisoning shows diffuse hypodensity of the cerebellum due to hypoxic-ischemic injury. The cerebrum shows normal density. The bilateral temporal horns of the lateral ventricles are dilated as a result of cerebellar swelling with compression of the 4th ventricle (A). Brain MRI on fluid attenuated inversion recovery (FLAIR) sequence performed 6 months later shows cortical necrosis with hyperintensity of the bilateral cerebellar hemispheres and sparing of the cerebrum and brain stem (B) and (C).

Patients with CO exposure in conventional magnetic resonance (MR) imaging may show increased T2-weighted hyper-intensities as compared with control [123]. Diffuse, symmetric white matter lesions, predominantly in the periventricular areas, although the centrum semiovale, deep subcortical white matter, thalamus, basal ganglia, and hippocampus also may be affected [123, 125-127] (Figure 2). New MR imaging techniques, such as diffusion tensor imaging (DTI) and apparent diffusion coefficient (ADC) may help to improve diagnostic accuracy and deepen our understanding of the pathophysiology of CO intoxication [84, 128].

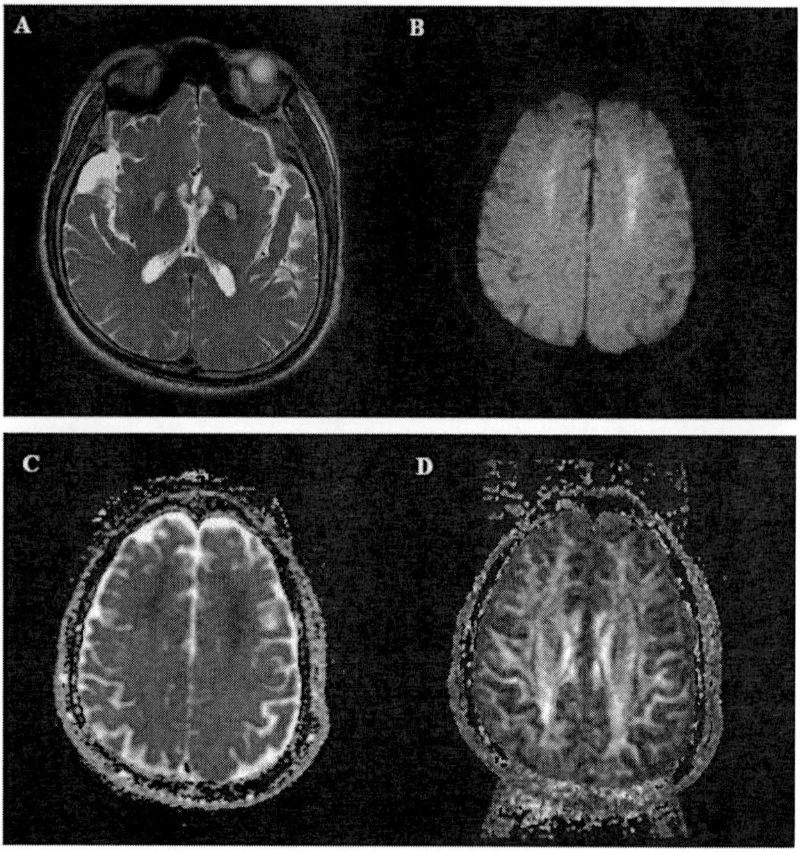

Figure 2. MR images of a 57-year-old man with delayed neuropsychological sequelae after carbon monoxide intoxication, showing high signal intensity of bilateral globus pallidi on an axial T2-weighted image (A), damage of cerebral white matter with restricted water diffusion of bilateral centrum semiovale shown as hyperintensity on axial diffusion-weighted image (B) and reciprocal hypointensity on axial apparent diffusion coefficient map (C), and reduced signal intensity of bilateral frontal white matter tract on fractional anisotropy map (D).

Fractional anisotropy (FA) and apparent diffusion coefficient (ADC) are quantitative indicators of DTI. The directionality of water diffusivity can be quantified by measuring fractional anisotropy (FA), whereas the apparent diffusion coefficient (ADC) indicates the degree of molecular mean diffusivity [84]. With these parameters, DTI provides quantitative information about the integrity of cerebral fiber tracts and may be a sensitive tool for identifying neuropathologic and neuropsychological changes associated with CO intoxication [84].

Moreover, lesions, such basal ganglia and hippocampus and other structures were identified years after poisoning (Figure 1), but DTI can show abnormal results one month after poisoning [129]. Although the severity in CO intoxication varied among individuals and could lead to different chronologic changes in FA and ADC values [129] and none of these abnormalities are specific to carbon monoxide poisoning, it is a consensus that patients are more likely to have poorer outcomes, such as death or persistent neurologic deficits when their neuroimaging findings are abnormal [78].

Neuropsychometric Testing

Several neuropsychometric tests has been designed specifically to screen for cognitive dysfunction as a result of CO poisoning [130], including, the Carbon Monoxide Neuropsychological Screening Battery (CONSB), Folstein Mini-Mental Status Examination (MMSE), Weschler adult intelligence scale–revised, Weschler memory scale–revised, and others [130].

CONSB consists of six subtests assessing general orientation, digit span, trail making, digit symbol, aphasia, and block design. Volunteers exposed to CO were found to perform poorly on the CONSB compared with controls without CO exposure [131]. CO-poisoned patients without concomitant drug and alcohol ingestion were found to score worse than controls before hyperbaric oxygen therapy (HBO) and to improve scores after HBO [130].

Glasgow Coma Scale (GCS) and the MMSE scores, quick and easy tests in clinical practices, may be potential predictors of the development of DNS [132]. However, motivation, practice effect due to repetition of the test, improvement of mental status overall, and metabolism of coingestants or cointoxicants may affect the results of the test [133-135]. Despite these limitations, neuropsychometric tests provide an objective means of evaluating cognitive function.

Treatment

Therapeutic strategies of the CO poisoning should include removal from the site of exposure, administration of supplemental oxygen, and general supportive care, including airway management, blood pressure support and stabilization of cardiovascular status. A suspected poisoned patient should administer normobaric oxygen by means of a face mask with a non-rebreather reservoir supplied with high-flow oxygen, or 100% oxygen, by means of an artificial airway, if appropriate. Poisoned patients should then be transported to an emergency department for further evaluation. If occult CO poisoning is suspected, however, other symptomatic cohabitants should be warned and evacuated until the source is identified and the environment is safe.

Whether oxygen should be given under normobaric oxygen (NBO) or increased pressure with hyperbaric oxygen therapy (HBO) is conflicting [54, 136]. A report, which is supportive of NBO, shows that patients presenting comatose after CO exposure had experienced a complete recovery even without the use of HBO [137]. Studies for comparing HBO with NBO for acute CO poisoning showed no beneficial [76] and significantly worse clinical symptoms and neuropsychometric test in the HBO group [138]. A Cochrane Review [139] collected seven trials [18, 69, 76, 138, 140-142] and found that pooled analysis did not suggest a benefit from HBO (OR for neurological deficits 0.78, 95%CI 0.54 to 1.12, $p=0.18$).

HBO consists of the delivery of 100% oxygen within a pressurized chamber resulting in a several-fold increase in the dissolved oxygen in the body. Hyperbaric oxygen therapy (HBOT) is defined as the breathing of 100% oxygen by patients within hyperbaric chambers compressed to greater than 1.4 atmospheres absolute (ATA)[143]. Although evidence based studies did not provide a strong conclusion, however, administration of NBO in a study did not show a reduction of cognitive sequelae after the inhalation of NBO, as compared with no supplemental oxygen therapy [81]. Additional two studies concluded that the incidence of DNS was zero in the HBO group and 12.1%[144] and 23% [69] in the NBO group. HBO reduces clinical recovery time and number of delayed functional abnormalities in non-comatose patients with acute CO poisoning [140]. HBO can improve outcome in patients with loss of consciousness, metabolic acidosis, CO-Hb level greater than 25%, and age older than 50[18]. In fact, HBO has been advocated to treat severe CO poisoning to limit delayed and permanent neurologic sequelae, because evidence shows that the administration of 100% oxygen at pressures higher than atmospheric further hastens the elimination of CO-Hb [145]. The effects of HBO are known to liberate hemoglobin by increasing the rate of dissociation of CO-Hb, allowed lung excretion of excessive CO, and accelerated the dissociation of CO from hemoglobin [146]. The half-life of CO-Hb is 240 to 320 minutes in a person at rest breathing room air (21% oxygen), 40 to 80 minutes at 100% oxygen, and approximately 20 minutes at 100% HBO 3 ATA [145, 147]. Therefore, increasing the partial pressure of oxygen can decrease the half-life of CO-Hb, result in more rapid dissociation of CO from respiratory cytochromes, and may antagonize the oxidative injury that occurs after CO poisoning. HBO directly reduces CO binding to hemoglobin [148] and cytochrome *c* oxidase [149], improves neutrophil adhesion to endothelium [150], decreases free radical production [150], reduces the severity of the neurologic deficit[151], enhances neuroprotection [152, 153], blocks CO-induced lipid peroxidation in brain [154], prevents the conversion of xanthine dehydrogenase to xanthine oxidase [155], and causes cerebral vasoconstriction, leading to reducing intracranial pressure and cerebral edema [156]. Most importantly, it can reduce overall mortality [157]. Together, HBO may have a role in preventing adverse neurologic sequelae in CO poisoning. The Undersea and Hyperbaric Medical Society recommends HBO for patients with transient or prolonged unconsciousness, presence of neurologic signs, cardiovascular dysfunction, or severe acidosis regardless of their CO-Hb levels, or patients who are 36 years of age or older, were exposed for 24 hours or more (including intermittent exposures), or have a carboxyhemoglobin level of 25% or more [143]. Pregnant women, particularly if the fetus shows signs of distress, should consider HBO firstly because of the increased toxicity risk to the fetus[158]. Patients with persistent symptoms should consider HBO despite NBO is used. If HBO therapy is not available, severe poisoning should be treated with 100% oxygen, with endotracheal intubation in patients who cannot protect their airway.

Therapeutic one session of HBO at 2.5 to 3.0 ATA is recommended initially, with further sessions considered if symptoms persist [134]. Patients not fulfilling criteria for HBO should receive 6 to 12 hours of 100% oxygen delivered by tight-fitting facemask [134, 138, 159, 160]. The HBO protocols for infants and children should be the same as that of adults[161]. HBO is recommended for pregnant patients with CO poisoning because of the potential benefit to the mother and unborn fetus [162]. The indications include a maternal CO-Hb level greater than 15% to 20% and evidence of fetal distress [158, 163] and a longer treatment with oxygen for pregnant patients is required [104, 164-167]. However, HBO is not entirely risk-free. Patients may complain of painful barotrauma affecting the ears and sinuses; patients with claustrophobia are often unable to tolerate the hyperbaric chamber. Other risks include oxygen toxicity, seizures, pulmonary edema and hemorrhage, decompression sickness including pneumothorax and nitrogen emboli, fire hazard [168, 169]. If a patient, who is in the process of HBO, requires emergency intervention, such as defibrillation, several minutes are required to decompress the patient safely before interventions can proceed.

Patients with mild poisoning can be treated in an ambulatory setting with high-flow oxygen. Patients with symptoms of moderate or severe CO poisoning, such as altered mental status or persistent neurologic or cardiovascular dysfunction may need to be admitted, because these patients may have comorbidities, such as concurrent cardiac ischemia, burns, or hemodynamic instability, each of which requires specialized care. Patients should be educated to minimize the risks of CO exposure, including by frequent inspection, maintenance and careful use of furnaces, gas stoves, and fireplaces. In addition, installation of CO detectors and avoidance of indoor unvented combustion sources such as grills and heaters may be possible to decrease the incidence of occult indoor CO poisoning. CO-Hb levels cannot reflect the true nature of an acute poisoning because the data depend on multiple factors, including the magnitude of exposure, the degree of alveolar ventilation, the blood volume, and metabolic activity, all causing CO-Hb levels poor correlation with clinical outcomes. In cases of loss of consciousness, arrhythmia and symptoms of cardiac insufficiency, CO intoxication should always be taken into consideration.

CONCLUSION

CO intoxication is the most notorious for its tastelessness, colorlessness, and odorlessness with variable sources of exposure. CO can bind rapidly to hemoglobin to form CO-Hb leading to tissue hypoxia. Patients may complain headache, nausea, vomiting, and flulike symptoms, or present further more serious manifestations, such as disturbed consciousness and sudden death. Failure to diagnose CO poisoning may result in significant morbidity and mortality and cause patients and cohabitants continued exposure to a dangerous environment. CO-Hb levels correlate poorly with clinical outcomes. The clues of CO poisoning are always from a careful history taking, astute physical examination, laboratory tests, brain imaging, and in particular, a high level of suspicion. Emergency treatment of CO poisoning begins with inhalation of supplemental oxygen and aggressive supportive care. CO poisoning can result in permanent sequelae.

Clinicians evaluating patients with acute poisoning should consider HBO because some randomized-trial data suggest a significant benefit of HBO. Rapid recognition and appropriate therapy with immediate supply of high dose oxygen may have a distinct reduction of mortality and long-term morbidity.

ACKNOWLEDGMENTS

This work was supported by the National Science Council, Taiwan, ROC (NSC99-2314-B-016-002-MY3 to S.-J. Chen) and Tri-Service General Hospital (TSGH-C100-024 to S.-J. Chen).

CONFLICT OF INTEREST

The authors have declared that no conflict of interest exists.

REFERENCES

[1] Hampson NB, Weaver LK: Carbon monoxide poisoning: a new incidence for an old disease. *Undersea Hyperb Med* 2007, 34:163-168.

[2] Peers C, Steele DS: Carbon monoxide: A vital signalling molecule and potent toxin in the myocardium. *J. Mol. Cell Cardiol* 2011. In press, corrected proof. http://www.sciencedirect.com/science/article/pii/S0022282811002148

[3] Kao LW, Nanagas KA: Carbon monoxide poisoning. *Med. Clin. North Am.* 2005, 89:1161-1194.

[4] Mah JC: Non-fire carbon monoxide deaths associated with the use of consumer produces: annual estimates:2007 annual estimates. Bethesda, MD: US Consumer Product Safety Commission. Available at: http://www.cpsc.gov/library/data.html. Accessed July 01, 2011.

[5] Wu MF, Huang SY: Indoor Carbon Monoxide Dilution and Alarm System with Wireless Device. *Asian J. Health and Information Sci* 2006, 1:228-236.

[6] Thom SR, Bhopale VM, Fisher D: Hyperbaric oxygen reduces delayed immune-mediated neuropathology in experimental carbon monoxide toxicity. *Toxicol Appl. Pharmacol* 2006, 213:152-159.
[7] Gorman D, Drewry A, Huang YL, Sames C: The clinical toxicology of carbon monoxide. *Toxicology* 2003, 187:25-38.
[8] Omaye ST: Metabolic modulation of carbon monoxide toxicity. *Toxicology* 2002, 180:139-150.
[9] Raub JA, Mathieu-Nolf M, Hampson NB, Thom SR: Carbon monoxide poisoning--a public health perspective. *Toxicology* 2000, 145:1-14.
[10] Goldbaum LR, Ramirez RG, Absalon KB: What is the mechanism of carbon monoxide toxicity? *Aviat Space Environ Med.* 1975, 46: 1289-1291.
[11] Grieb G, Groger A, Bozkurt A, Stoffels I, Piatkowski A, Pallua N: The diversity of carbon monoxide intoxication: medical courses can differ extremely-a case report. *Inhal. Toxicol.* 2008, 20:911-915.
[12] Harper A, Croft-Baker J: Carbon monoxide poisoning: undetected by both patients and their doctors. *Age Ageing* 2004, 33:105-109.
[13] Weaver LK: Clinical practice. Carbon monoxide poisoning. *N. Engl. J. Med.* 2009, 360:1217-1225.
[14] Silver S, Smith C, Worster A: Should hyperbaric oxygen be used for carbon monoxide poisoning? *CJEM* 2006, 8:43-46.
[15] Prockop LD, Chichkova RI: Carbon monoxide intoxication: an updated review. *J Neurol Sci* 2007, 262:122-130.
[16] Stoller KP: Hyperbaric oxygen and carbon monoxide poisoning: a critical review. *Neurol Res* 2007, 29:146-155.
[17] Hampson NB, Dunford RG, Kramer CC, Norkool DM: Selection criteria utilized for hyperbaric oxygen treatment of carbon monoxide poisoning. *J. Emerg Med* 1995, 13:227-231.
[18] Weaver LK, Hopkins RO, Chan KJ, Churchill S, Elliott CG, Clemmer TP, Orme JF, Jr., Thomas FO, Morris AH: Hyperbaric oxygen for acute carbon monoxide poisoning. *N. Engl. J. Med.* 2002, 347:1057-1067.
[19] Ely EW, Moorehead B, Haponik EF: Warehouse workers' headache: emergency evaluation and management of 30 patients with carbon monoxide poisoning. *Am. J. Med.* 1995, 98:145-155.
[20] Thom SR, Bhopale VM, Han ST, Clark JM, Hardy KR: Intravascular neutrophil activation due to carbon monoxide poisoning. *Am. J. Respir. Crit. Care Med.* 2006, 174:1239-1248.

[21] Piantadosi CA, Zhang J, Levin ED, Folz RJ, Schmechel DE: Apoptosis and delayed neuronal damage after carbon monoxide poisoning in the rat. *Exp. Neurol.* 1997, 147:103-114.

[22] Shimada T, Yamazaki H, Mimura M, Inui Y, Guengerich FP: Interindividual variations in human liver cytochrome P-450 enzymes involved in the oxidation of drugs, carcinogens and toxic chemicals: studies with liver microsomes of 30 Japanese and 30 Caucasians. *J. Pharmacol. Exp. Ther* 1994, 270:414-423.

[23] Omura T, Sato R: The carbon monoxide-binding pigment of liver microsomes. II. Solubilization, purification, and properities. *J. Biol. Chem.* 1964, 239:2379-2385.

[24] Omura T, Sato R: The carbon monoxide-binding pigment of liver microsomes. I. Evidence for its hemoprotein nature. *J. Biol. Chem.* 1964, 239:2370-2378.

[25] Yao P, Hao L, Nussler N, Lehmann A, Song F, Zhao J, Neuhaus P, Liu L, Nussler A: The protective role of HO-1 and its generated products (CO, bilirubin, and Fe) in ethanol-induced human hepatocyte damage. *Am. J. Physiol Gastrointest Liver Physiol.* 2009, 296:G1318-G1323.

[26] Ordway GA, Garry DJ: Myoglobin: an essential hemoprotein in striated muscle. *J. Exp. Biol.* 2004, 207:3441-3446.

[27] Pyriochou A, Papapctropoulos A: Soluble guanylyl cyclase: more secrets revealed. *Cell Signal* 2005, 17:407-413.

[28] Ignarro LJ: Heme-dependent activation of soluble guanylate cyclase by nitric oxide: regulation of enzyme activity by porphyrins and metalloporphyrins. *Semin Hematol* 1989, 26:63-76.

[29] Behrends S: Drugs that activate specific nitric oxide sensitive guanylyl cyclase isoforms independent of nitric oxide release. *Curr. Med. Chem.* 2003, 10:291-301.

[30] Evgenov OV, Kohane DS, Bloch KD, Stasch JP, Volpato GP, Bellas E, Evgenov NV, Buys ES, Gnoth MJ, Graveline AR, Liu R, Hess DR, Langer R, Zapol WM: Inhaled agonists of soluble guanylate cyclase induce selective pulmonary vasodilation. *Am. J. Respir. Crit. Care Med.* 2007, 176:1138-1145.

[31] Martin E, Berka V, Bogatenkova E, Murad F, Tsai AL: Ligand selectivity of soluble guanylyl cyclase: effect of the hydrogen-bonding tyrosine in the distal heme pocket on binding of oxygen, nitric oxide, and carbon monoxide. *J. Biol. Chem* 2006, 281:27836-27845.

[32] Marks GS, Brien JF, Nakatsu K, McLaughlin BE: Does carbon monoxide have a physiological function? *Trends Pharmacol. Sci.* 1991, 12:185-188.

[33] Barinaga M: Carbon monoxide: killer to brain messenger in one step. *Science* 1993, 259:309.

[34] Verma A, Hirsch DJ, Glatt CE, Ronnett GV, Snyder SH: Carbon monoxide: a putative neural messenger. *Science* 1993, 259:381-384.

[35] Landry DW, Oliver JA: The pathogenesis of vasodilatory shock. *N Engl J Med* 2001, 345:588-595.

[36] Jiang J, Tyssebotn I: Cerebrospinal fluid pressure changes after acute carbon monoxide poisoning and therapeutic effects of normobaric and hyperbaric oxygen in conscious rats. *Undersea Hyperb. Med.* 1997, 24:245-254.

[37] Meilin S, Rogatsky GG, Thom SR, Zarchin N, Guggenheimer-Furman E, Mayevsky A: Effects of carbon monoxide on the brain may be mediated by nitric oxide. *J. Appl. Physiol.* 1996, 81:1078-1083.

[38] Thom SR, Fisher D, Manevich Y: Roles for platelet-activating factor and *NO-derived oxidants causing neutrophil adherence after CO poisoning. *Am. J. Physiol Heart Circ. Physiol.* 2001, 281:H923-H930.

[39] Thom SR, Ohnishi ST, Ischiropoulos H: Nitric oxide released by platelets inhibits neutrophil B2 integrin function following acute carbon monoxide poisoning. *Toxicol. Appl. Pharmacol.* 1994, 128:105-110.

[40] Hardy KR, Thom SR: Pathophysiology and treatment of carbon monoxide poisoning. *J. Toxicol. Clin. Toxicol.* 1994, 32:613-629.

[41] Thom SR: Carbon monoxide-mediated brain lipid peroxidation in the rat. *J. Appl. Physiol.* 1990, 68:997-1003.

[42] Thom SR: Leukocytes in carbon monoxide-mediated brain oxidative injury. *Toxicol Appl. Pharmacol* 1993, 123:234-247.

[43] Ischiropoulos H, Beers MF, Ohnishi ST, Fisher D, Garner SE, Thom SR: Nitric oxide production and perivascular nitration in brain after carbon monoxide poisoning in the rat. *J. Clin. Invest* 1996, 97: 2260-2267.

[44] Thom SR, Garner S, Fisher D, Ischiropoulos H: Vascular nitrosative stress from CO exposure. *Undersea Hyperb Med* 1998, 25:47.

[45] Zhang J, Piantadosi CA: Mitochondrial oxidative stress after carbon monoxide hypoxia in the rat brain. *J Clin Invest* 1992, 90:1193-1199.

[46] Brunelle JK, Bell EL, Quesada NM, Vercauteren K, Tiranti V, Zeviani M, Scarpulla RC, Chandel NS: Oxygen sensing requires mitochondrial ROS but not oxidative phosphorylation. *Cell Metab* 2005, 1:409-414.

[47] Moncada S, Erusalimsky JD: Does nitric oxide modulate mitochondrial energy generation and apoptosis? *Nat. Rev. Mol. Cell Biol.* 2002, 3: 214-220.

[48] Kulisz A, Chen N, Chandel NS, Shao Z, Schumacker PT: Mitochondrial ROS initiate phosphorylation of p38 MAP kinase during hypoxia in cardiomyocytes. *Am. J. Physiol. Lung Cell Mol. Physiol* 2002, 282:L1324-L1329.

[49] Mehta SR, Niyogi M, Kasthuri AS, Dubal U, Bindra S, Prasad D, Lahiri AK: Carbon monoxide poisoning. *J. Assoc Physicians India* 2001, 49:622-625.

[50] Hara S, Mukai T, Kurosaki K, Kuriiwa F, Endo T: Characterization of suppression of nitric oxide production by carbon monoxide poisoning in the striatum of free-moving rats, as determined by in vivo brain microdialysis. *Brain Res.* 2003, 979:27-36.

[51] Penney DG: Hemodynamic response to carbon monoxide. *Environ. Health Perspect.* 1988, 77:121-130.

[52] Yildirim C, Gunay N, Buyukaslan H, Kucukdurmaz Z, Bozkurt S: A case of carbon monoxide poisoning with thrombus in the heart: a case report. *Inhal Toxicol.* 2005, 17:797-801.

[53] Onvlee-Dekker IM, De Vries AC, Ten Harkel AD: Carbon monoxide poisoning mimicking long-QT induced syncope. *Arch. Dis. Child* 2007, 92:244-245.

[54] Hampson NB, Zmaeff JL: Outcome of patients experiencing cardiac arrest with carbon monoxide poisoning treated with hyperbaric oxygen. *Ann. Emerg. Med.* 2001, 38:36-41.

[55] Marius-Nunez AL: Myocardial infarction with normal coronary arteries after acute exposure to carbon monoxide. *Chest* 1990, 97:491-494.

[56] Koehler RC, Jones MD, Jr., Traystman RJ: Cerebral circulatory response to carbon monoxide and hypoxic hypoxia in the lamb. *Am. J. Physiol.* 1982, 243:H27-H32.

[57] Song SY, Okeda R, Funata N, Higashino F: An experimental study of the pathogenesis of the selective lesion of the globus pallidus in acute carbon monoxide poisoning in cats. With special reference to the chronologic change in the cerebral local blood flow. *Acta Neuropathol* 1983, 61:232-238.

[58] Ginsberg MD, Myers RE, McDonagh BF: Experimental carbon monoxide encephalopathy in the primate. II. Clinical aspects, neuropathology, and physiologic correlation. *Arch Neurol* 1974, 30: 209-216.

[59] Olson KR: Carbon monoxide poisoning: mechanisms, presentation, and controversies in management. *J. Emerg. Med.* 1984, 1:233-243.

[60] Okeda R, Funata N, Takano T, Miyazaki Y, Higashino F, Yokoyama K, Manabe M: The pathogenesis of carbon monoxide encephalopathy in the acute phase--physiological and morphological correlation. *Acta Neuropathol.* 1981, 54:1-10.

[61] Myers RA, Snyder SK, Majerus TC: Cutaneous blisters and carbon monoxide poisoning. *Ann. Emerg. Med.* 1985, 14:603-606.

[62] Florkowski CM, Rossi ML, Carey MP, Poulton K, Dickson GR, Ferner RE: Rhabdomyolysis and acute renal failure following carbon monoxide poisoning: two case reports with muscle histopathology and enzyme activities. *J. Toxicol. Clin. Toxicol.* 1992, 30:443-454.

[63] Herman GD, Shapiro AB, Leikin J: Myonecrosis in carbon monoxide poisoning. *Vet Hum Toxicol* 1988, 30:28-30.

[64] Wolff E: Carbon monoxide poisoning with severe myonecrosis and acute renal failure. *Am. J. Emerg. Med.* 1994, 12:347-349.

[65] Thom SR: Smoke inhalation. *Emerg. Med. Clin. North Am.* 1989, 7: 371-387.

[66] Krantz T, Thisted B, Strom J, Sorensen MB: Acute carbon monoxide poisoning. *Acta Anaesthesiol. Scand* 1988, 32:278-282.

[67] Grieb G, Simons D, Piatkowski A, Altiok E, Eppstein RJ, Bernhagen J, Pallua N: Carbon monoxide intoxication versus myocardial infarction: An easy diagnosis? *Burns* 2011, 37:e29-e31.

[68] Watanabe S, Matsuo H, Kobayashi Y, Satoh Y, Fujita M, Nakayama A, Aizawa Y, Shinomiya N, Suzuki S: Transient degradation of myelin basic protein in the rat hippocampus following acute carbon monoxide poisoning. *Neurosci. Res.* 2010, 68:232-240.

[69] Thom SR, Taber RL, Mendiguren II, Clark JM, Hardy KR, Fisher AB: Delayed neuropsychologic sequelae after carbon monoxide poisoning: prevention by treatment with hyperbaric oxygen. *Ann. Emerg. Med.* 1995, 25:474-480.

[70] Lee MS, Marsden CD: Neurological sequelae following carbon monoxide poisoning clinical course and outcome according to the clinical types and brain computed tomography scan findings. *Mov. Disord* 1994, 9:550-558.

[71] Thom SR, Keim LW: Carbon monoxide poisoning: a review epidemiology, pathophysiology, clinical findings, and treatment options including hyperbaric oxygen therapy. *J. Toxicol. Clin. Toxicol.* 1989, 27:141-156.

[72] Min SK: A brain syndrome associated with delayed neuropsychiatric sequelae following acute carbon monoxide intoxication. *Acta Psychiatr Scand* 1986, 73:80-86.

[73] Myers RA: Carbon monoxide poisoning. *J. Emerg. Med.* 1984, 1: 245-248.

[74] Choi IS: Delayed neurologic sequelae in carbon monoxide intoxication. *Arch. Neurol.* 1983, 40:433-435.

[75] Norkool DM, Kirkpatrick JN: Treatment of acute carbon monoxide poisoning with hyperbaric oxygen: a review of 115 cases. *Ann. Emerg. Med.* 1985, 14:1168-1171.

[76] Raphael JC, Elkharrat D, Jars-Guincestre MC, Chastang C, Chasles V, Vercken JB, Gajdos P: Trial of normobaric and hyperbaric oxygen for acute carbon monoxide intoxication. *Lancet* 1989, 2:414-419.

[77] Smith JS, Brandon S: Morbidity from acute carbon monoxide poisoning at three-year follow-up. *Br. Med. J.* 1973, 1:318-321.

[78] Jones JS, Lagasse J, Zimmerman G: Computed tomographic findings after acute carbon monoxide poisoning. *Am. J. Emerg. Med.* 1994, 12:448-451.

[79] Myers RA, Britten JS: Are arterial blood gases of value in treatment decisions for carbon monoxide poisoning? *Crit. Care Med.* 1989, 17:139-142.

[80] Choi IS, Kim SK, Lee SS, Choi YC: Evaluation of outcome of delayed neurologic sequelae after carbon monoxide poisoning by technetium-99m hexamethylpropylene amine oxime brain single photon emission computed tomography. *Eur. Neurol.* 1995, 35:137-142.

[81] Weaver LK, Valentine KJ, Hopkins RO: Carbon monoxide poisoning: risk factors for cognitive sequelae and the role of hyperbaric oxygen. *Am J. Respir. Crit Care Med.* 2007, 176:491-497.

[82] Tibbles PM, Perrotta PL: Treatment of carbon monoxide poisoning: a critical review of human outcome studies comparing normobaric oxygen with hyperbaric oxygen. *Ann. Emerg. Med.* 1994, 24:269-276.

[83] Beppu T, Nishimoto H, Ishigaki D, Fujiwara S, Yoshida T, Oikawa H, Kamada K, Sasaki M, Ogasawara K: Assessment of damage to cerebral white matter fiber in the subacute phase after carbon monoxide poisoning using fractional anisotropy in diffusion tensor imaging. *Neuroradiology* 2010, 52:735-743.

[84] Lin WC, Lu CH, Lee YC, Wang HC, Lui CC, Cheng YF, Chang HW, Shih YT, Lin CP: White matter damage in carbon monoxide intoxication

assessed in vivo using diffusion tensor MR imaging. *AJNR Am. J. Neuroradiol.* 2009, 30:1248-1255.

[85] Chu K, Jung KH, Kim HJ, Jeong SW, Kang DW, Roh JK: Diffusion-weighted MRI and 99mTc-HMPAO SPECT in delayed relapsing type of carbon monoxide poisoning: evidence of delayed cytotoxic edema. *Eur. Neurol.* 2004, 51:98-103.

[86] Hon KL, Yeung WL, Ho CH, Leung WK, Li AM, Chu WC, Chan YL: Neurologic and radiologic manifestations of three girls surviving acute carbon monoxide poisoning. *J. Child Neurol.* 2006, 21:737-741.

[87] Choi IS, Lee MS, Lee YJ, Kim JH, Lee SS, Kim WT: Technetium-99m HM-PAO SPECT in patients with delayed neurologic sequelae after carbon monoxide poisoning. *J. Korean Med. Sci.* 1992, 7:11-18.

[88] Hopkins RO, Weaver LK: Cognitive outcomes 6 years after acute carbon monoxide poisoning. *Undersea Hyperb. Med.* 2008, 35:258.

[89] Weaver LK, Hopkins RO, hurchill S, Deru K: Neurological outcomes 6 years after acute carbon monoxide poisoning. *Undersea Hyperb. Med.* 2008, 35:258-259.

[90] Penney DG: Chronic carbon monoxide poisoning: a case series. In *Carbon monoxide poisoning*. Edited by Penney DG: Boca Raton, FL: CRC Press; 2008:551-567.

[91] Fawcett TA, Moon RE, Fracica PJ, Mebane GY, Theil DR, Piantadosi CA: Warehouse workers' headache. Carbon monoxide poisoning from propane-fueled forklifts. *J. Occup. Med.* 1992, 34:12-15.

[92] Johnson BL, Cohen HH, Struble R, Setzer JV, Anger WK, Gutnik BD, McDonough T, Hauser P: Behavioral Toxicology: Early Detection of Occupational Hazards.: Washington, DC, US Government Printing Office; 1974:306-328.

[93] Ryan CM: Memory disturbances following chronic, low-level carbon monoxide exposure. *Arch. Clin. Neuropsychol.* 1990, 5:59-67.

[94] Atkins EH, Baker EL: Exacerbation of coronary artery disease by occupational carbon monoxide exposure: a report to two fatalities and a review of the literature. *Am. J. Ind. Med.* 1985, 7:73-79.

[95] Myers RA, DeFazio A, Kelly MP: Chronic carbon monoxide exposure: a clinical syndrome detected by neuropsychological tests. *J. Clin. Psychol.* 1998, 54:555-567.

[96] Liebelt EL: Hyperbaric oxygen therapy in childhood carbon monoxide poisoning. *Curr. Opin. Pediatr.* 1999, 11:259-264.

[97] Crocker PJ, Walker JS: Pediatric carbon monoxide toxicity. *J. Emerg. Med.* 1985, 3:443-448.

[98] Foster M, Goodwin SR, Williams C, Loeffler J: Recurrent acute life-threatening events and lactic acidosis caused by chronic carbon monoxide poisoning in an infant. *Pediatrics* 1999, 104:e34.

[99] Vreman HJ, Mahoney JJ, Stevenson DK: Carbon monoxide and carboxyhemoglobin. *Adv. Pediatr.* 1995, 42:303-334.

[100] Elkharrat D, Raphael JC, Korach JM, Jars-Guincestre MC, Chastang C, Harboun C, Gajdos P: Acute carbon monoxide intoxication and hyperbaric oxygen in pregnancy. *Intensive Care Med.* 1991, 17:289-292.

[101] Caravati EM, Adams CJ, Joyce SM, Schafer NC: Fetal toxicity associated with maternal carbon monoxide poisoning. *Ann. Emerg. Med.* 1988, 17:714-717.

[102] Woody RC, Brewster MA: Telencephalic dysgenesis associated with presumptive maternal carbon monoxide intoxication in the first trimester of pregnancy. *J. Toxicol. Clin. Toxicol.* 1990, 28:467-475.

[103] Koren G, Sharav T, Pastuszak A, Garrettson LK, Hill K, Samson I, Rorem M, King A, Dolgin JE: A multicenter, prospective study of fetal outcome following accidental carbon monoxide poisoning in pregnancy. *Reprod. Toxicol.* 1991, 5:397-403.

[104] Farrow JR, Davis GJ, Roy TM, McCloud LC, Nichols GR: Fetal death due to nonlethal maternal carbon monoxide poisoning. *J. Forensic. Sci.* 1990, 35:1448-1452.

[105] Cramer CR: Fetal death due to accidental maternal carbon monoxide poisoning. *J. Toxicol. Clin. Toxicol.* 1982, 19:297-301.

[106] Ginsberg MD, Myers RE: Fetal brain injury after maternal carbon monoxide intoxication. Clinical and neuropathologic aspects. *Neurology* 1976, 26:15-23.

[107] Davis SM, Levy RC: High carboxyhemoglobin level without acute or chronic findings. *J. Emerg. Med.* 1984, 1:539-542.

[108] Burney RE, Wu SC, Nemiroff MJ: Mass carbon monoxide poisoning: clinical effects and results of treatment in 184 victims. *Ann. Emerg. Med.* 1982, 11:394-399.

[109] Piantadosi CA: Diagnosis and treatment of carbon monoxide poisoning. *Respir Care Clin. N Am* 1999, 5:183-202.

[110] Ernst A, Zibrak JD: Carbon monoxide poisoning. *N. Engl. J. Med.* 1998, 339:1603-1608.

[111] Stewart RD, Baretta ED, Platte LR, Stewart EB, Kalbfleisch JH, Van YB, Rimm AA: Carboxyhemoglobin levels in American blood donors. *JAMA* 1974, 229:1187-1195.

[112] Touger M, Gallagher EJ, Tyrell J: Relationship between venous and arterial carboxyhemoglobin levels in patients with suspected carbon monoxide poisoning. *Ann. Emerg. Med.* 1995, 25:481-483.

[113] Widdop B: Analysis of carbon monoxide. *Ann. Clin. Biochem.* 2002, 39:378-391.

[114] Buckley RG, Aks SE, Eshom JL, Rydman R, Schaider J, Shayne P: The pulse oximetry gap in carbon monoxide intoxication. *Ann. Emerg. Med.* 1994, 24:252-255.

[115] Bozeman WP, Myers RA, Barish RA: Confirmation of the pulse oximetry gap in carbon monoxide poisoning. *Ann. Emerg. Med.* 1997, 30:608-611.

[116] Hampson NB: Pulse oximetry in severe carbon monoxide poisoning. *Chest* 1998, 114:1036-1041.

[117] Turner M, Esaw M, Clark RJ: Carbon monoxide poisoning treated with hyperbaric oxygen: metabolic acidosis as a predictor of treatment requirements. *J. Accid. Emerg. Med.* 1999, 16:96-98.

[118] Sokal JA, Kralkowska E: The relationship between exposure duration, carboxyhemoglobin, blood glucose, pyruvate and lactate and the severity of intoxication in 39 cases of acute carbon monoxide poisoning in man. *Arch. Toxicol.* 1985, 57:196-199.

[119] Silver DA, Cross M, Fox B, Paxton RM: Computed tomography of the brain in acute carbon monoxide poisoning. *Clin Radiol* 1996, 51: 480-483.

[120] Halavaara J, Valanne L, Setala K: Neuroimaging supports the clinical diagnosis of methanol poisoning. *Neuroradiology* 2002, 44:924-928.

[121] Inoue N: [Extrapyramidal syndrome induced by chemical substances]. *Nippon Rinsho* 1993, 51:2924-2928.

[122] Hart IK, Kennedy PG, Adams JH, Cunningham NE: Neurological manifestation of carbon monoxide poisoning. *Postgrad Med. J.* 1988, 64:213-216.

[123] Parkinson RB, Hopkins RO, Cleavinger HB, Weaver LK, Victoroff J, Foley JF, Bigler ED: White matter hyperintensities and neuropsychological outcome following carbon monoxide poisoning. *Neurology* 2002, 58:1525-1532.

[124] Fan HC, Wang AC, Lo CP, Chang KP, Chen SJ: Damage of cerebellar white matter due to carbon monoxide poisoning: a case report. *Am. J. Emerg. Med.* 2009, 27:757.

[125] Durak AC, Coskun A, Yikilmaz A, Erdogan F, Mavili E, Guven M: Magnetic resonance imaging findings in chronic carbon monoxide intoxication. *Acta Radiol.* 2005, 46:322-327.

[126] Pavese N, Napolitano A, De IG, Canapicchi R, Collavoli PL, Lucetti C, Gambaccini G, Bonuccelli U: Clinical outcome and magnetic resonance imaging of carbon monoxide intoxication. A long-term follow-up study. *Ital. J. Neurol. Sci.* 1999, 20:171-178.

[127] Pulsipher DT, Hopkins RO, Weaver LK: Basal ganglia volumes following CO poisoning: A prospective longitudinal study. *Undersea Hyperb Med* 2006, 33:245-256.

[128] Murata T, Kimura H, Kado H, Omori M, Onizuka J, Takahashi T, Itoh H, Wada Y: Neuronal damage in the interval form of CO poisoning determined by serial diffusion weighted magnetic resonance imaging plus 1H-magnetic resonance spectroscopy. *J. Neurol. Neurosurg. Psychiatry* 2001, 71:250-253.

[129] Terajima K, Igarashi H, Hirose M, Matsuzawa H, Nishizawa M, Nakada T: Serial assessments of delayed encephalopathy after carbon monoxide poisoning using magnetic resonance spectroscopy and diffusion tensor imaging on 3.0T system. *Eur. Neurol.* 2008, 59:55-61.

[130] Messier LD, Myers RA: A neuropsychological screening battery for emergency assessment of carbon-monoxide-poisoned patients. *J. Clin. Psychol.* 1991, 47:675-684.

[131] Amitai Y, Zlotogorski Z, Golan-Katzav V, Wexler A, Gross D: Neuropsychological impairment from acute low-level exposure to carbon monoxide. *Arch. Neurol.* 1998, 55:845-848.

[132] Ku HL, Yang KC, Lee YC, Lee MB, Chou YH: Predictors of carbon monoxide poisoning-induced delayed neuropsychological sequelae. *Gen. Hosp. Psychiatry* 2010, 32:310-314.

[133] Deschamps D, Geraud C, Julien H, Baud FJ, Dally S: Memory one month after acute carbon monoxide intoxication: a prospective study. *Occup. Environ. Med.* 2003, 60:212-216.

[134] Hampson NB, Mathieu D, Piantadosi CA, Thom SR, Weaver LK: Carbon monoxide poisoning: interpretation of randomized clinical trials and unresolved treatment issues. *Undersea Hyperb. Med.* 2001, 28: 157-164.

[135] Schiltz KL: Failure to assess motivation, need to consider psychiatric variables, and absence of comprehensive examination: a skeptical review of neuropsychologic assessment in carbon monoxide research. *Undersea Hyperb.Med.* 2000, 27:48-50.

[136] Gilmer B, Kilkenny J, Tomaszewski C, Watts JA: Hyperbaric oxygen does not prevent neurologic sequelae after carbon monoxide poisoning. *Acad. Emerg. Med.* 2002, 9:1-8.

[137] Weaver LK, Hopkins RO, Larson-Lohr V: Neuropsychologic and functional recovery from severe carbon monoxide poisoning without hyperbaric oxygen therapy. *Ann. Emerg. Med.* 1996, 27:736-740.

[138] Scheinkestel CD, Bailey M, Myles PS, Jones K, Cooper DJ, Millar IL, Tuxen DV: Hyperbaric or normobaric oxygen for acute carbon monoxide poisoning: a randomised controlled clinical trial. *Med. J. Aust* 1999, 170:203-210.

[139] Juurlink DN, Buckley NA, Stanbrook MB, Isbister GK, Bennett M, McGuigan MA: Hyperbaric oxygen for carbon monoxide poisoning. *Cochrane Database Syst. Rev.* 2005:CD002041.

[140] Ducasse JL, Celsis P, Marc-Vergnes JP: Non-comatose patients with acute carbon monoxide poisoning: hyperbaric or normobaric oxygenation? *Undersea Hyperb. Med.* 1995, 22:9-15.

[141] Managing carbon monoxide poisoning with hyperbaric oxygen. In *J. Toxicol. Clin. Toxicol.* 2004, 42:455-456.

[142] Randomized prospective study comparing the effect of HBO vs. 12 hours NBO in non-comatose CO-poisoned patients: results of the preliminary analysis. In *Undersea Hyperb. Med.* 1996, 23 Suppl:7.

[143] Gesell LB: *Hyperbaric oxygen 2009: indications and results: the Hyperbaric Oxygen Therapy Committee report. Durham, NC: Undersea and Hyperbaric Medical Society.*; 2008.

[144] Myers RA, Snyder SK, Emhoff TA: Subacute sequelae of carbon monoxide poisoning. *Ann. Emerg. Med.* 1985, 14:1163-1167.

[145] Jay GD, McKindley DS: Alterations in pharmacokinetics of carboxyhemoglobin produced by oxygen under pressure. *Undersea Hyperb. Med.* 1997, 24:165-173.

[146] Smith G: The treatment of carbon monoxide poisoning with oxygen at two atmospheres absolute. *Ann. Occup. Hyg.* 1962, 5:259-263.

[147] Ilano AL, Raffin TA: Management of carbon monoxide poisoning. *Chest* 1990, 97:165-169.

[148] Jay GD, McKindley DS: Carboxyhemoglobin toxicokinetics. *Acad. Emerg. Med.* 1999, 6:766-768.

[149] Brown SD, Piantadosi CA: Reversal of carbon monoxide-cytochrome c oxidase binding by hyperbaric oxygen in vivo. *Adv Exp Med Biol* 1989, 248:747-754.

[150] Thom SR: Functional inhibition of leukocyte B2 integrins by hyperbaric oxygen in carbon monoxide-mediated brain injury in rats. *Toxicol. Appl. Pharmacol.* 1993, 123:248-256.

[151] Katsenelson K, Arieli R, Arieli Y, Abramovich A, Feinsod M, Tal D: Hyperbaric oxygen pretreatment according to the gas micronuclei denucleation hypothesis reduces neurologic deficit in decompression sickness in rats. *J. Appl. Physiol.* 2009, 107:558-563.

[152] Li JS, Zhang W, Kang ZM, Ding SJ, Liu WW, Zhang JH, Guan YT, Sun XJ: Hyperbaric oxygen preconditioning reduces ischemia-reperfusion injury by inhibition of apoptosis via mitochondrial pathway in rat brain. *Neuroscience* 2009, 159:1309-1315.

[153] Wada K, Miyazawa T, Nomura N, Tsuzuki N, Nawashiro H, Shima K: Preferential conditions for and possible mechanisms of induction of ischemic tolerance by repeated hyperbaric oxygenation in gerbil hippocampus. *Neurosurgery* 2001, 49:160-166.

[154] Thom SR: Antagonism of carbon monoxide-mediated brain lipid peroxidation by hyperbaric oxygen. *Toxicol. Appl. Pharmacol.* 1990, 105:340-344.

[155] Brown SD, Piantadosi CA: Recovery of energy metabolism in rat brain after carbon monoxide hypoxia. *J. Clin. Invest* 1992, 89:666-672.

[156] Sukoff MH, Ragatz RE: Hyperbaric oxygenation for the treatment of acute cerebral edema. *Neurosurgery* 1982, 10:29-38.

[157] Mannaioni PF, Vannacci A, Masini E: Carbon monoxide: the bad and the good side of the coin, from neuronal death to anti-inflammatory activity. *Inflamm. Res.* 2006, 55:261-273.

[158] Van Hoesen KB, Camporesi EM, Moon RE, Hage ML, Piantadosi CA: Should hyperbaric oxygen be used to treat the pregnant patient for acute carbon monoxide poisoning? A case report and literature review. *JAMA* 1989, 261:1039-1043.

[159] Finnerty JP: Hyperbaric oxygen for acute carbon monoxide poisoning. *N Engl. J. Med.* 2003, 348:557-560.

[160] Piantadosi CA: Carbon monoxide poisoning. *N. Engl. J. Med.* 2002, 347:1054-1055.

[161] Rudge FW: Carbon monoxide poisoning in infants: treatment with hyperbaric oxygen. *South Med. J.* 1993, 86:334-337.

[162] Bar R, Cohen M, Bentur Y, Shupak A, Adir Y: Pre-Labor exposure to carbon monoxide: should the neonate be treated with hyperbaric oxygenation? *Clin. Toxicol. (Phila)* 2007, 45:579-581.

[163] Tomaszewski CA, Thom SR: Use of hyperbaric oxygen in toxicology. *Emerg Med Clin North Am* 1994, 12:437-459.

[164] Brown DB, Mueller GL, Golich FC: Hyperbaric oxygen treatment for carbon monoxide poisoning in pregnancy: a case report. *Aviat Space Environ Med* 1992, 63:1011-1014.

[165] Longo LD: Carbon monoxide in the pregnant mother and fetus and its exchange across the placenta. *Ann N Y Acad Sci* 1970, 174:312-341.

[166] Longo LD, Hill EP: Carbon monoxide uptake and elimination in fetal and maternal sheep. *Am. J. Physiol.* 1977, 232:H324-H330.

[167] Margulies JL: Acute carbon monoxide poisoning during pregnancy. *Am J. Emerg. Med.* 1986, 4:516-519.

[168] Hampson NB, Simonson SG, Kramer CC, Piantadosi CA: Central nervous system oxygen toxicity during hyperbaric treatment of patients with carbon monoxide poisoning. *Undersea Hyperb. Med.* 1996, 23: 215-219.

[169] Weaver LK, Hopkins RO, Elliott G: Carbon monoxide poisoning. *N. Engl. J. Med.* 1999, 340:1290.

[170] Rajiah K, Mathew EM: Clinical manifestation, effects, diagnosis, monitoring of carbon monoxide poisoning and toxicity. *Afr. J. Pharm. Pharmacol.* 2011, 5:259-264.

[171] Mathieu D, Mathieu-Nolf M, Linke J, Favory R, Wattel F: *Handbook on Hyperbaric Medicine*: *Springer*; 2006.

In: Carbon Monoxide: Sources, Uses and Hazards ISBN:978-1-61942-055-7
Editors: D. DiLoreto et al.

Chapter 3

CO OXIDATION USING GOLD SUPPORTED ON CE-MN-O COMPOSITE MATERIALS

S. A. C. Carabineiro[a], A. M. T. Silva[a], G. Dražič[b], P. B. Tavares[c], and J. L. Figueiredo[a]

[a]Laboratório de Catálise e Materiais (LCM), Laboratório Associado LSRE/LCM, Faculdade de Engenharia, Universidade do Porto, Porto, Portugal

[b]Jozef Stefan Institute, Department of Nanostructured Materials, Ljubljana, Slovenia

[c]Universidade de Trás-os-Montes e Alto Douro, CQVR Centro de Química Departamento de Química, Vila Real, Portugal

ABSTRACT

Carbon monoxide (CO) is a colourless, odourless, highly toxic, poisonous gas. Exposure to a few hundreds ppm can cause permanent health damages or even death. Low level exposure, if continued, can also cause health problems in the long term. CO can thus be considered as one of the most dangerous environmental pollutants. Therefore, its removal and detection is of extreme importance for safe living.

Catalysts which promote CO oxidation at room temperature will find many practical applications that include air-purification devices for indoor space (including gas masks), and for cleaning automotive exhaust gases. CO removal is also required in CO_2 lasers, and in polymer electrolyte fuel cell (PEFC) usage, where the CO concentration in the hydrogen rich gas must be selectively reduced to less than 100 ppm.

Haruta and co-workers have intensively investigated the ability of gold catalysts to catalyse this reaction. In their early work in the 80s, the activity of supported gold catalysts came as a surprise, but it is now well known that supported gold nanocatalysts are effective for CO oxidation at very low temperatures, being active even at -70°C. However, activities depend on the choice of the metal oxide support.

In the present work, Ce-Mn-O supports with different Ce/Mn ratios were prepared by co-precipitation, in addition to cerium oxide only and manganese oxide only. These supports were loaded with gold by a double impregnation method. Samples were characterised by adsorption of N_2 at -196°C, temperature programmed reduction (TPR), X-ray diffraction (XRD), high-resolution transmission electron microscopy (HRTEM) and high-angle annular dark-field imaging (HAADF) with Z-contrast. CO oxidation was studied using a continuous-flow reactor connected to a gas chromatograph.

It was found that the higher the amount of ceria in the sample, the most active it was for CO oxidation. The ability to convert CO seems to be independent of the surface area of the supports, but seems to vary according to the temperature of the first reduction peak of the TPR spectrum, in the case of samples loaded with gold. Also it varies according to the gold nanoparticle size.

INTRODUCTION

Carbon monoxide (CO) is a colourless, odourless, highly toxic, poisonous gas. It can be formed due to photochemical reactions in the troposphere. Other natural sources of CO include volcanoes, forest fires, and other forms of combustion. Exposure to a few hundreds ppm can cause permanent health damages or even death. Low level exposure, if continued, can also cause health problems in the long term. CO can thus be considered as one of the most dangerous environmental pollutants. Therefore, its removal and detection is of extreme importance for safe living.

CO oxidation ($CO + \frac{1}{2} O_2 \rightarrow CO_2$) is very simple, but this reaction, that has many potential applications in fuel cells and sensing, is not fully understood [1-11]. Catalysts which promote this reaction will find many practical applications that include air-purification devices for indoor space (including gas masks), and cleaning automotive exhaust gases [1, 2, 9-11]. CO removal is also required in CO_2 lasers, and in polymer electrolyte fuel cells

(PEFC), where the CO concentration in the hydrogen rich gas must be selectively reduced to less than 100 ppm [10].

Haruta and co-workers have intensively investigated the ability of gold catalysts to catalyse this reaction [3-8, 12-19]. In their early work in the 80s, the activity of supported gold catalysts came as a surprise, but it is now well known that supported gold nanocatalysts are effective for CO oxidation at very low temperatures, being active even at -70°C [10, 11]. Moreover, it is well known from the literature that gold is essentially inert in the bulk state, and that it is only active in the form of nanoparticles well dispersed on the support [3-8, 10-28].

Regarding the CO oxidation mechanism, although it has been extensively studied in several catalystic systems, it is still under debate [2, 9-11, 20-22, 25, 27]. In fact, for gold on an oxide support, the Bond and Thompson mechanism (Figure 1), initially proposed by these authors in 2000 [2] and substantiated by subsequent results of several other authors [29], is widely accepted.

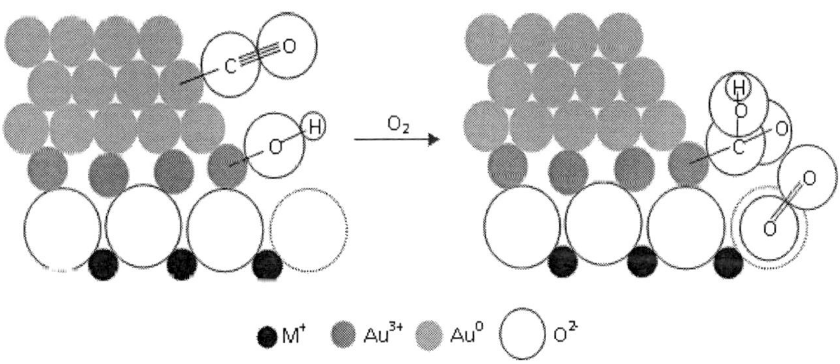

Figure 1. Representation of the early stages of the oxidation of carbon monoxide at the periphery of an active gold particle. At the left, a carbon monoxide molecule is chemisorbed on a low coordination gold atom, and an hydroxyl ion moves from the support to an Au^{3+} ion, creating an anion vacancy. At the right they have reacted to form a carboxylate group, and an oxygen molecule occupies the anion vacancy as O_2^-. This then oxidizes the carboxylate group by extracting an hydrogen atom, forming carbon dioxide, and the resulting hydroperoxide ion HO_2^- then oxidizes a further carboxylate species forming another carbon dioxide and restoring two hydroxide ions to the support surface. This completes the catalytic cycle. No attempt is made to suggest charges carried by the reacting species (redrawn from [11]).

Exceptions are, naturally, some “inert” supports like alumina and silica, where oxygen chemisorption will be negligible and involvement of lattice oxide ions is unlikely; in consequence, the mechanism on gold particles, where there is no support participation and where the total length of the periphery is comparatively small, will be of the Langmuir–Hinshelwood type, involving oxygen (weakly) and carbon monoxide (more strongly) adsorbed adjacently onto gold [2]. Au/ceria catalysts have shown to be very efficient in CO oxidation [21, 26-28, 30-68], but manganese oxides are quite active, with or without gold [20, 50, 69-78]. Ce-Mn-O materials have been reported in the literature [79-83] and used in reactions such as wet oxidation [74, 84-96], alcohol oxidation [97-99], VOC abatement [100, 101], fluoride removal from water [102], NOx reduction or oxidation [103-107], ozonation [108, 109] or hydrocarbon combustion [110-115]. However, to the best of our knowledge, they were never used in CO oxidation (only when doped with Co for preferential oxidation of CO in the presence of H_2 (PROX) [116-118]). As far as we know, there are very few reports on Au/Ce-Mn-O materials in the literature, which consist in catalysts prepared by incipient wetness impregnation of CeO_2 with a manganese oxide precursor, for various Ce/Mn ratios ($Ce_{90}Mn_{10}$ and $Ce_{50}Mn_{50}$) [119], or co-precipitation (CP) with a Ce/Mn ratio of 1:1 [120, 121], which were also used in the PROX reaction. In all cases, Au was loaded onto the supports by deposition-precipitation (DP) [119-121], an improvement being reported when the support was ultrasonically dispersed before Au loading [120, 121].

Therefore, to the best of our knowledge there are no reports on the use of Ce-Mn-O composites prepared by co-precipitation, with Au loaded by a double impregnation method (DIM) for CO oxidation, and this was the purpose of the present work.

Experimental

Ce-Mn-O supports with different Ce/Mn ratios (hereafter denoted as Ce_xMn_y, x and y being the molar percentages of Ce and Mn, respectively) were prepared by co-precipitation (CP) [88-90], by mixing aqueous solutions of the corresponding metal nitrates as precursors, namely manganese nitrate (Sigma Aldrich) and cerium nitrate (Sigma Aldrich). The mixture was poured into 200 mL of a 3 M NaOH solution, the precipitate was filtrated and washed several times with ultra pure water, dried over night at 100°C and finally calcinated at 300°C. The molar composition ratio was 30%, 50% or 70% of the active metal (Ce or Mn). The following materials were obtained $Ce_{30}Mn_{70}$, $Ce_{50}Mn_{50}$, $Ce_{70}Mn_{30}$. In addition, cerium oxide (Ce_{100}) and manganese oxide (Mn_{100}) were also synthesised using the same procedure. After preparation, all materials were crushed into fine powders. These supports were loaded with gold by a double impregnation method [20, 21, 24, 26, 27, 122], which briefly consists in impregnation with a solution of $HAuCl_4$ (using sonication), followed by addition of an aqueous solution of Na_2CO_3 (1 M), under constant ultrasonic stirring. The resulting solid was washed repeatedly with distilled water for chloride removal (which is well known to cause sinterization of Au nanoparticles, thus turning them inactive [1, 10, 11, 28, 122]), and dried in the oven at ~110°C overnight. Samples were characterised by adsorption of N_2 at -196°C, temperature programmed reduction (TPR), X-ray diffraction (XRD), high-resolution transmission electron microscopy (HRTEM) and Z-contrast imaging using high-angle annular dark-field (HAADF) detector. Chemical composition was determined with energy-dispersive X-ray spectroscopy (EDXS).

CO oxidation was studied using a continuous-flow reactor (0.2 g, 5% CO, 10% O_2 in He, 50 ml/min) connected to a gas chromatograph. Further details can be found elsewhere [20, 21, 27].

RESULTS AND DISCUSSION

BET Surface Area

Table 1 shows the BET surface areas of the Ce-Mn-O supports. The area of Ce_{100} is more than 3 times larger than that of Mn_{100}. Both the areas of Ce_{100} and Mn_{100} increase by addition of Mn or Ce, respectively, with a maximum for $Ce_{50}Mn_{50}$.

Table 1. BET surface areas of the Ce-Mn-O samples obtained by adsorption of N_2 at 77K, crystallite sizes and phases detected by XRD

Sample	BET area (m^2/g)	Phase identified (crystal system, space group)
Ce_{100}	83	CeO_2 (cerianite, cubic, Fm-3m)
$Ce_{70}Mn_{30}$	140	CeO_2 (cerianite) + Mn_3O_4 (hausmanite)
$Ce_{50}Mn_{50}$	147	CeO_2 (cerianite) + Mn_3O_4 (hausmanite)
$Ce_{30}Mn_{70}$	116	CeO_2 (cerianite) + Mn_3O_4 (hausmanite)
Mn_{100}	24	Mn_3O_4 (hausmanite, tetragonal , I41/amd)

This is in contrast with other authors that report an increase in the areas with Ce content increase, for samples prepared by DP [119]. However, Qi et al. [104] reported that mixed oxides prepared by CP (the same method used in this work) have the highest surface area with a manganese ratio of 0.4.

This shows that the surface area of Ce-Mn-O composites is influenced by the preparation method. Chen et al. [86] reported that the surface area increased from 9 to 77 m^2/g with an increase in CeO_2 content from 0 to 15% for samples prepared by CP.

XRD

Table 1 and Figure 2 shows the XRD results obtained. CeO_2 (cerianite, cubic, Fm-3m) was detected on Ce_{100}, while Mn_3O_4 (hausmanite, tetragonal, I41/amd) was found on Mn_{100}. Other samples showed a mixture of these phases, as expected. Other authors obtained CeO_2 and β-MnO_2 [119]. Other manganese oxides phases such as α-Mn_2O_3, β-MnO_2, and Mn_5O_8, for materials prepared by CP of Mn(II) and Ce(III) chloride aqueous solutions, have also been reported [86, 91]. The particle size of CeO_2 and Mn_3O_4 were around 11 and 33 nm, respectively. Smaller values were found for these phases on the Ce-Mn-O composite materials, independently of the Ce/Mn ratio, as also found by other authors [119]. Gold was not detected by XRD, possibly due to the low loading.

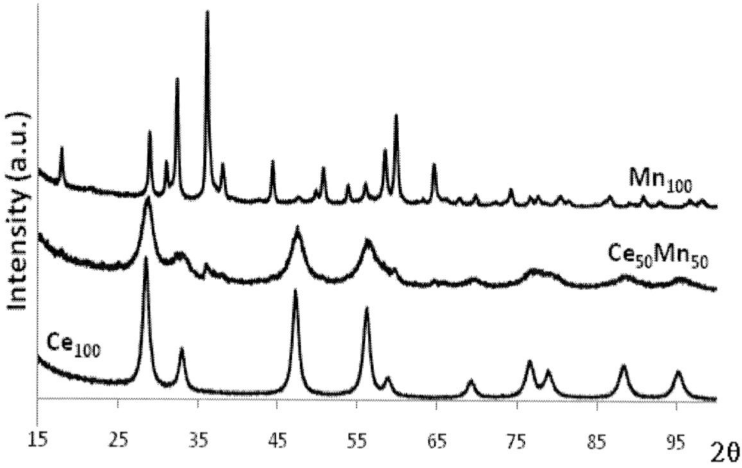

Figure 2. X-ray diffraction spectra of Ce-Mn-O supports.

TPR

TPR results of Ce-Mn-O supports are shown in Figure 3. Two peaks are seen in the spectrum of Ce100, corresponding to the reduction of surface oxygen (~450ºC) and bulk oxygen (~800ºC), as referred in the literature [21, 24, 25, 27, 28, 121]. The two peaks observed in the Mn_{100} spectrum are from the reduction of Mn_3O_4 to Mn_2O_3 (~300ºC) and of the latter to MnO (~450ºC), as also referred in the literature [20, 101, 114, 121]. The characteristic green color of the sample obtained at the end of the reduction analysis proves the transformation into MnO [20]. The first peak is not seen on the $Ce_{70}Mn_{30}$, but appears in $Ce_{50}Mn_{50}$. As the amount of Mn added increases, the intensity of the second peak increases.

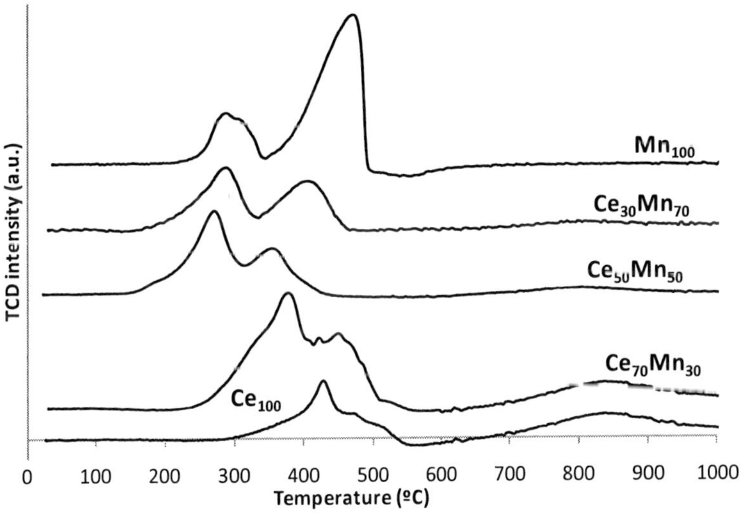

Figure 3. TPR spectra of Ce-Mn-O supports.

Loading gold shifts the ceria surface oxygen reduction peak to much lower temperatures (~100ºC) as seen in Figure 4, showing that the presence of Au helped to improve the reducibility of the surface oxygen [21, 24, 25, 27, 28, 121]. As the amount of Ce decreases the shift is less intense and the peak is no longer visible in the $Ce_{30}Mn_{70}$ material. Not much shift was observed in the manganese oxide peaks upon gold loading. Meng et al. referred that, after supporting 1 wt% Au, the reduction of Mn oxides and partial ceria surface capping oxygen are promoted, possibly due to the hydrogen spillover effect [121].

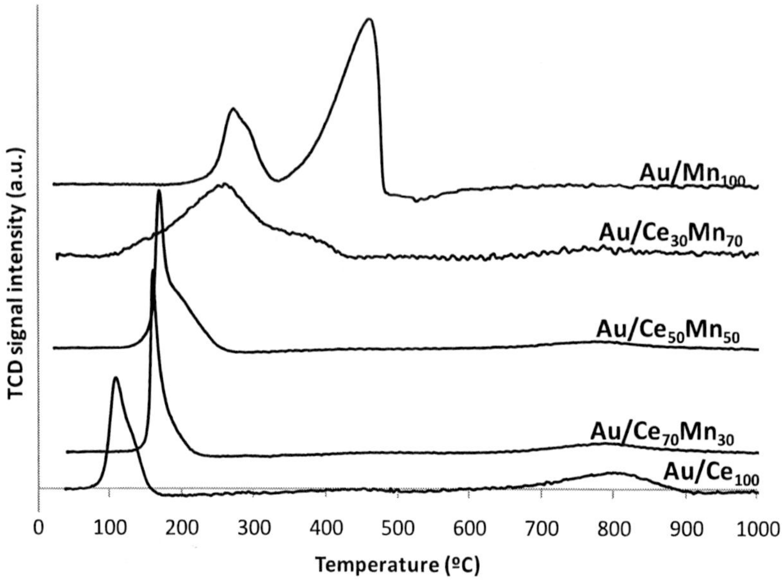

Figure 4. TPR spectra of Au/Ce-Mn-O samples.

HRTEM, HAADF and EDXS

Figures 5 and 6 show HRTEM, HAADF and EDXS results. Gold particles are seen as darker spots in HRTEM and as bright dots in HAADF images. Particle sizes are displayed in Table 2.

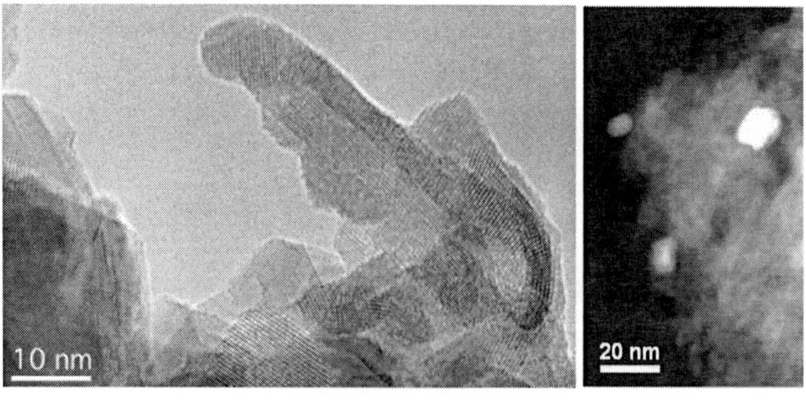

Figure 5. HRTEM image of the $Ce_{30}Mn_{70}$ support (left) and HAADF image of $Au/Ce_{30}Mn_{70}$ material (right).

Table 2. Average sizes and ranges for Au nanoparticles on the Au/Ce-Mn-O samples

Au/Ce-Mn-O sample	Gold size range (nm)	Gold size average (nm)
Au/Ce_{100}	2-9	5.1
Au/$Ce_{70}Mn_{30}$	2-16	8.6
Au/$Ce_{50}Mn_{50}$	4-18	11.4
Au/$Ce_{30}Mn_{70}$	7-20	12.8
Au/Mn_{100}	> 20	> 20

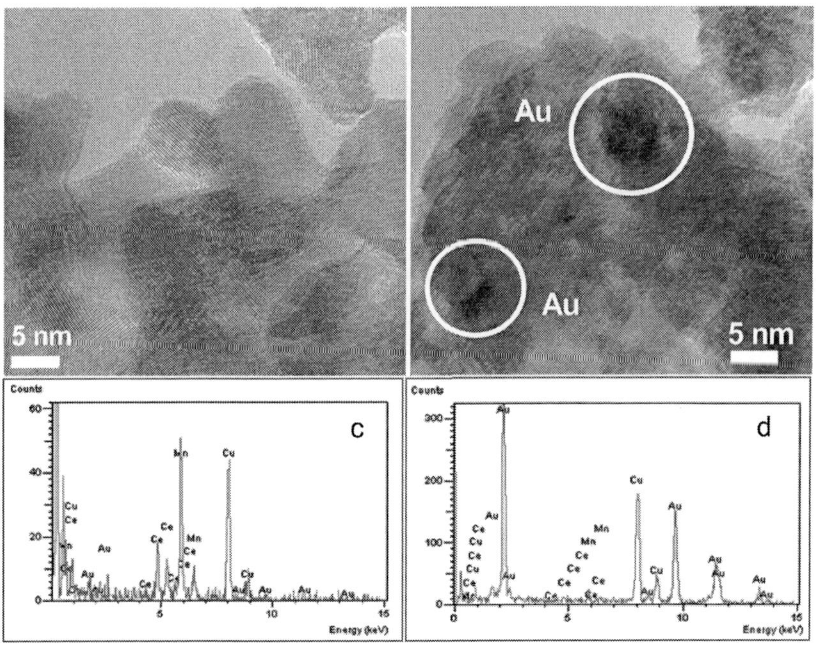

Figure 6. HRTEM images of the $Ce_{70}Mn_{30}$ support alone (top, left) and loaded with Au (top, right), with respective EDXS spectra taken from an average of grains (bottom, left) and from a gold nanoparticle (bottom, right). The Cu peaks are from the carbon-coated Cu grid used to support samples.

The Au nanoparticle sizes are much smaller on Ce_{100} (2-9 nm, average 5.1 nm) than on Mn_{100} (> 20 nm). In fact, addition of Ce to Mn promotes a particle size decrease. Other authors also obtained larger Au NP sizes in Mn_{100} than in Ce_{100} [119].

CO Oxidation Results

Figures 7 and 8 show the CO conversion results for the Ce-Mn-O samples with and without Au. The ability to convert CO seems to be independent of the surface area of the supports, but varying accordingly with the amount of ceria present in the sample (Figure 7). Similar results were obtained with Ce-Ti-O materials [7].

Addition of gold to the supports increases up to 15 times the catalytic activity at room temperature. With the unloaded samples, full conversion was only achieved above 350°C; but when Au was loaded, full CO conversion could be achieved already at 150°C (for Au/Ce_{100} and Au/$Ce_{70}Mn_{30}$). The same trend was maintained when Au was loaded, that is, the catalytic activity varies with the amount of ceria present (Figure 8). Also it varies according to the gold nanoparticle size that, as said above, decreases with the increase of ceria amount in the support.

It can be observed that, for Ce-Mn-O materials, the temperature at which the CO conversion starts (Figure 7) is closely related with the temperature of the first peak of the TPR spectrum (Figure 3).

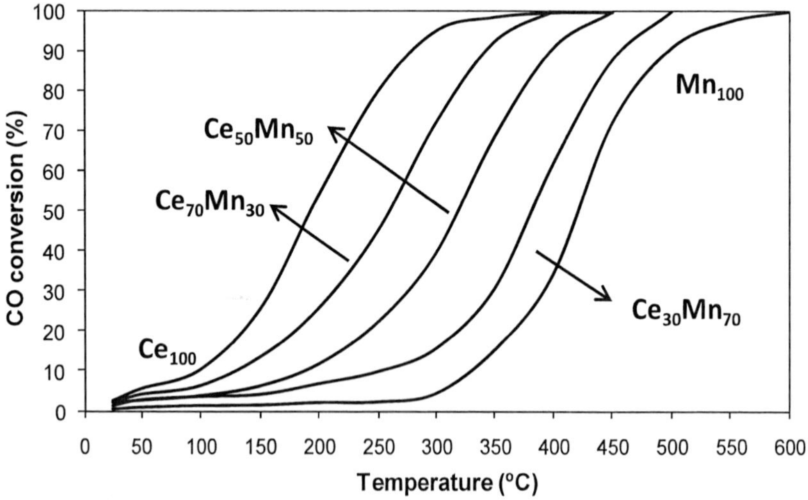

Figure 7. CO conversion results (%) versus temperature for the Ce-Mn-O samples.

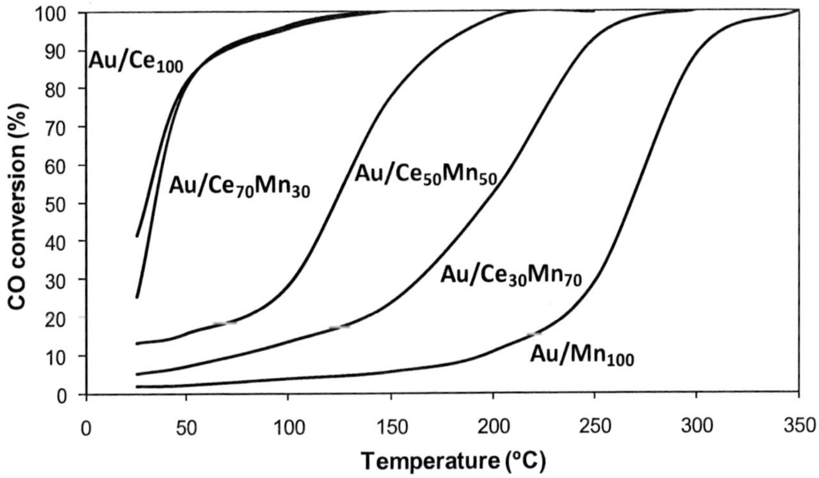

Figure 8. CO conversion results (%) versus temperature for the Au/Ce-Mn-O samples.

The lower the temperature of the first TPR peak, the more active is the catalyst. It is generally accepted that CO oxidation under stationary conditions occurs over ceria by a Mars-van Krevelen type mechanism, where reaction involves alternate reduction and oxidation of the ceria surface with formation of surface oxygen vacancies (as the key step) and their successive replenishment by gas-phase oxygen, the formation and desorption of CO_2 completing the cycle [21]. The obtained results seem to confirm that this mechanism takes place also with the Ce-Mn-O catalysts.

As for samples loaded with Au, there is also a relationship with the temperature of the first TPR peak of the TPR spectrum (Figure 4) with the temperature at which the CO conversion starts (Figure 8). This again shows the importance of the lattice oxygen in the catalytic activity, suggesting that the Bond and Thompson mechanism [2, 29], described in the Introduction, is also taking place with this kind of materials.

CONCLUSIONS

Ce-Mn-O supports with different Ce/Mn ratios were prepared by CP. The BET surface area of Ce_{100} and Mn_{100} increase by addition of Mn or Ce, respectively, with a maximum for $Ce_{50}Mn_{50}$. The area of Ce_{100} is more than 3 times larger than that of Mn_{100}. CeO_2 was detected on Ce_{100}, while Mn_3O_4 was found on Mn_{100}. Ce-Mn-O samples showed a mixture of these phases, as expected.

TPR results of Ce-Mn-O supports showed two peaks at the spectrum of Ce_{100}, corresponding to the reduction of surface oxygen (~450°C) and bulk oxygen (~800°C). The two peaks observed in the Mn_{100} spectrum are from the reduction of Mn_3O_4 to Mn_2O_3 (~300°C) and of the latter to MnO (~450°C). The first peak is not seen on the $Ce_{70}Mn_{30}$, but appears in $Ce_{50}Mn_{50}$. The intensity of the second peak increases with the amount of Mn added. Loading gold shifts the ceria surface oxygen reduction peak to much lower temperatures (~100°C), showing that the presence of Au helped to improve the reducibility of the surface oxygen. Not much shift was observed in the manganese oxide peaks upon gold loading.

The Au nanoparticle sizes are much smaller on Ce_{100} (2-9 nm, average 5.1 nm) than on Mn_{100} (> 20 nm). The addition of Ce to Mn promotes a particle size decrease.

The ability to convert CO seems to be independent of the surface area of the supports, but varies accordingly with the amount of ceria present in the sample. Addition of gold to the supports increases up to 15 times the catalytic activity at room temperature and decreased the temperature for maximum conversion. The catalytic activity of the gold loaded samples varies with the amount of ceria present and with the gold nanoparticle size, that decreases with the increase of ceria amount in the support.

The temperature at which CO conversion starts is closely related with the temperature of the first peak of the TPR spectrum, showing the importance of the lattice oxygen in the catalytic activity. This suggests that the Bond and Thompson mechanism is taking place with the gold loaded samples.

ACKNOWLEDGMENTS

Authors are grateful to Fundação para a Ciência e a Tecnologia (Portugal) and the Ministry of Higher Education, Science and Technology (Slovenia) for financial support (Portugal-Slovenia Cooperation in Science and Technology 2010-2011). SAC and AMTS also acknowledge FCT for CIENCIA 2007 program and POCI/N010/2006, respectively. GD acknowledges the financial support of the Slovenian Research Agency.

REFERENCES

[1] Bond, GC; Thompson, DT. Catalysis by gold. *Catalysis Reviews-Science and Engineering. 1*999, 41, 319-388.

[2] Bond, GC; Thompson, DT. Gold-catalysed oxidation of carbon monoxide. *Gold Bulletin*. 2000, 33, 41-51.

[3] Haruta, M. Size- and support-dependency in the catalysis of noble metals: Gold as a typical example. *Physics and Chemistry of Clusters*. 2001, 117, 99-109.

[4] Haruta, M; Date, M. Advances in the catalysis of Au nanoparticles. *Applied Catalysis a-General*. 2001, 222, 427-437.

[5] Haruta, M. Catalysis of gold nanoparticles deposited on metal oxides. *Cattech*. 2002, 6, 102-115.

[6] Haruta, A. When gold is not noble: Catalysis by nanoparticles. *Chemical Record*. 2003, 3, 75-87.

[7] Haruta, M. Catalysis and applications of gold nanoparticles. Science and *Technology in Catalysis* 2002. 2003, 145, 31-38.

[8] Haruta, M. Nanoparticulate gold catalysts for low-temperature CO oxidation. *Journal of New Materials for Electrochemical Systems*. 2004, 7, 163-172.

[9] Bond, GC; Louis, C; Thompson, DT. *Catalysis by Gold*. Hutchings, GJ, editor. vol. 6. London, United Kingdom: Imperial College Press; 2006.

[10] Carabineiro, SAC; Thompson, DT. Catalytic Applications for Gold Nanotechnology. In: Heiz, EU, Landman, U, editors. *Nanocatalysis*. vol. Berlin, Heidelberg, New York: Springer-Verlag; 2007. p. 377-489.

[11] Carabineiro, SAC; Thompson, DT. Gold Catalysis. In: Corti, C, Holliday, R, editors. *Gold: Science and Applications*. vol. Boca Raton, London, New York: CRC Press, Taylor and Francis Group; 2010. p. 89-122.

[12] Haruta, M; Kobayashi, T; Sano, H; Yamada, N. Novel gold catalysts for the oxidation of carbon monoxide at a temperature far below 0°C. *Chemistry Letters*. 1987, 405-408.

[13] Haruta, M; Yamada, N; Kobayashi, T; Iijima, S. Gold catalysts prepared by coprecipitation for low-temperature oxidation of hydrogen and carbon-monoxide. *Journal of Catalysis*. 1989, 115, 301-309.

[14] Haruta, M; Tsubota, S; Kobayashi, T; Kageyama, H; Genet, MJ; Delmon, B. Low-temperature oxidation of CO over gold supported on TiO_2, alpha-Fe_2O_3 and Co_3O_4. *Journal of Catalysis*. 1993, 144, 175-192.

[15] Haruta, M; Tsubota, S; Kobayashi, T; Ueda, A; Sakurai, H; Ando, M. Mechanistic studies of CO oxidation on highly dispersed gold catalysts for use in room-temperature air purification. *New Frontiers in Catalysis*, Pt C. 1993, 75, 2657-2660.

[16] Haruta, M; Tsubota, S; Ueda, A; Sakurai, H. Synergism in the catalysis of supported gold. *New Aspects of Spillover Effect in Catalysis*. 1993, 77, 45-52.

[17] Haruta, M. Size- and support-dependency in the catalysis of gold. *Catalysis Today*. 1997, 36, 153-166.

[18] Haruta, M. Gold as a low-temperature oxidation catalyst: factors controlling activity and selectivity. 3rd World Congress on Oxidation *Catalysis*. 1997, 110, 123-134.

[19] Haruta, M. Gold as a novel catalyst in the 21st century: Preparation, working mechanism and applications. *Gold Bulletin*. 2004, 37, 27-36.

[20] Carabineiro, SAC; Bastos, SST; Órfão, JJM; Pereira, MFR; Delgado, JJ; Figueiredo, JL. Carbon monoxide oxidation catalysed by exotemplated manganese oxides. *Catalysis Letters*. 2010, 134, 217-227.

[21] Carabineiro, SAC; Bastos, SST; Órfão, JJM; Pereira, MFR; Delgado, JJ; Figueiredo, JL. Exotemplated ceria catalysts with gold for CO oxidation. *Applied Catalysis A: General*. 2010, 381, 150-160.

[22] Carabineiro, SAC; Machado, BF; Bacsa, RR; Serp, P; Dražić, G; Faria, JL; Figueiredo, JL. Catalytic performance of Au/ZnO nanocatalysts for CO oxidation. *Journal of Catalysis*. 2010, 273, 191–198.

[23] Carabineiro, SAC; Machado, BF; Drazic, G; Bacsa, RR; Serp, P; Figueiredo, JL; Faria, JL. Photodeposition of Au and Pt on ZnO and TiO_2. In: Gaigneaux, EM, Devillers, M, Hermans, S, Jacobs, PA, Martens, JA, Ruiz, P, editors. *Studies in Surface Science and Catalysis*. vol. 175: Elsevier; 2010. p. 629-633.

[24] Carabineiro, SAC; Silva, AMT; Dražić, G; Figueiredo, JL. Preparation of Au nanoparticles on Ce-Ti-O supports. In: Gaigneaux, EM, Devillers, M, Hermans, S, Jacobs, PA, Martens, JA, Ruiz, P, editors. *Studies in Surface Science and Catalysis*. vol. 175: Elsevier; 2010. p. 457-461.

[25] Carabineiro, SAC; Bogdanchikova, N; Avalos-Borja, M; Pestryakov, A; Tavares, P; Figueiredo, JL. Gold supported on metal oxides for carbon monoxide oxidation. *Nano Research*. 2011, 4, 180–193.

[26] Carabineiro, SAC; Silva, AMT; Dražić, G; Figueiredo, JL. Analytical electron microscopy of gold nanoparticles on ceria, titania and ceria-titania materials. In: Méndez-Vilas, A, Díaz, J, editors. Microscopy: Science, Technology, Applications and Education, *Formatex Microscopy Book series*. vol. 4: Formatex Research Center; 2010. p. in press.

[27] Carabineiro, SAC; Silva, AMT; Dražić, G; Tavares, PB; Figueiredo, JL. Gold nanoparticles on ceria supports for the oxidation of carbon monoxide. *Catalysis today*. 2010, 154, 21–30.

[28] Carabineiro, SAC; Silva, AMT; Dražić, G; Tavares, PB; Figueiredo, JL. Effect of chloride on the sinterization of Au/CeO_2 catalysts. *Catalysis Today*. 2010, 154, 293-302.

[29] Bond, G; Thompson, D. Formulation of mechanisms for gold-catalysed reactions. *Gold Bulletin*. 2009, 42, 247-259.

[30] Bera, P; Hegde, MS. Characterization and catalytic properties of combustion synthesized Au/CeO_2 catalyst. *Catalysis Letters*. 2002, 79, 75-81.

[31] Carrettin, S; Concepcion, P; Corma, A; Nieto, JML; Puntes, VF. Nanocrystalline CeO_2 increases the activity of an for CO oxidation by two orders of magnitude. *Angewandte Chemie-International Edition*. 2004, 43, 2538-2540.

[32] Glaspell, G; Fuoco, L; El-Shall, MS. Microwave synthesis of supported Au and Pd nanoparticle catalysts for CO oxidation. *Journal of Physical Chemistry B*. 2005, 109, 17350-17355.

[33] Guzman, J; Carrettin, S; Corma, A. Spectroscopic evidence for the supply of reactive oxygen during CO oxidation catalyzed by gold supported on nanocrystalline CeO_2. *Journal of the American Chemical Society*. 2005, 127, 3286-3287.

[34] Guzman, J; Carrettin, S; Fierro-Gonzalez, JC; Hao, YL; Gates, BC; Corma, A. CO oxidation catalyzed by supported gold: Cooperation between gold and nanocrystalline rare-earth supports forms reactive surface superoxide and peroxide species. *Angewandte Chemie-International Edition*. 2005, 44, 4778-4781.

[35] Guzman, J; Carrettin, S; Corma, A. Spectroscopic evidence of the supply of reactive oxygen during CO oxidation catalysed by gold supported on nanocrystalline CeO_2. *Gold Bulletin*. 2005, 38, 133-133.

[36] Huang, PX; Wu, F; Zhu, BL; Gao, XP; Zhu, HY; Yan, TY; Huang, WP; Wu, SH; Song, DY. CeO_2 nanorods and gold nanocrystals supported on

CeO_2 nanorods as catalyst. *Journal of Physical Chemistry* B. 2005, 109, 19169-19174.

[37] Manzoli, M; Chiorino, A; Boccuzzi, F. Nanostructure and catalytic properties of gold supported on group IV oxides. In: Gamble, ACCCS, editor. *Oxide Based Materials*: *New Sources*, Novel Phases, New Applications. vol. 1552005. p. 405-413.

[38] Venezia, AM; Pantaleo, G; Longo, A; Di Carlo, G; Casaletto, MP; Liotta, FL; Deganello, G. Relationship between structure and CO oxidation activity of ceria-supported gold catalysts. *Journal of Physical Chemistry B*. 2005, 109, 2821-2827.

[39] Arena, F; Famulari, P; Interdonato, N; Bonura, G; Frusteri, F; Spadaro, L. Physico-chemical properties and reactivity of Au/CeO_2 catalysts in total and selective oxidation of COD. *Catalysis Today*. 2006, 116, 384-390.

[40] Arena, F; Famulari, P; Trunfio, G; Bonura, G; Frusteri, F; Spadaro, L. Probing the factors affecting structure and activity of the Au/CeO_2 system in total and preferential oxidation of CO. *Applied Catalysis B-Environmental*. 2006, 66, 81-91.

[41] Concepcion, P; Carrettin, S; Corma, A. Stabilization of cationic gold species on Au/CeO_2 catalysts under working conditions. *Applied Catalysis a-General*. 2006, 307, 42-45.

[42] Centeno, MA; Hadjiivanov, K; Venkov, T; Klimev, H; Odriozola, JA. Comparative study of Au/Al_2O_3 and Au/CeO_2-Al_2O_3 catalysts. *Journal of Molecular Catalysis a-Chemical*. 2006, 252, 142-149.

[43] Ivanova, S; Pitchon, V; Petit, C. Application of the direct exchange method in the preparation of gold catalysts supported on different oxide materials. *Journal of Molecular Catalysis a-Chemical*. 2006, 256, 278-283.

[44] Lai, SY; Qiu, YF; Wang, SJ. Effects of the structure of ceria on the activity of gold/ceria catalysts for the oxidation of carbon monoxide and benzene. *Journal of Catalysis*. 2006, 237, 303-313.

[45] Moreau, F; Bond, GC. CO oxidation activity of gold catalysts supported on various oxides and their improvement by inclusion of an iron component. *Catalysis Today*. 2006, 114, 362-368.

[46] Park, ED. CO oxidation over Au/CeO_2 prepared by a co-precipitation method. In: Rhee, HKNISPJM, editor. New Development and Application in Chemical Reaction Engineering, 4th Asia-Pacific *Chemical Reaction Engineering Symposium*. vol. 1592006. p. 321-324.

[47] Pillai, UR; Deevi, S. Highly active gold-ceria catalyst for the room temperature oxidation of carbon monoxide. *Applied Catalysis a-General*. 2006, 299, 266-273.

[48] Solsona, BE; Garcia, T; Jones, C; Taylor, SH; Carley, AF; Hutchings, GJ. Supported gold catalysts for the total oxidation of alkanes and carbon monoxide. *Applied Catalysis a-General*. 2006, 312, 67-76.

[49] Wang, X-Y; Wang, S-P; Wang, S-R; Zhao, Y-Q; Huang, J; Zhang, S-M; Huang, W-P; Wu, S-H. The preparation of Au/CeO_2 catalysts and their activities for low-temperature CO oxidation. *Catalysis Letters*. 2006, 112, 115-119.

[50] Chang, L-H; Sasirekha, N; Rajesh, B; Chen, Y-W. CO oxidation on ceria- and manganese oxide-supported gold catalysts. *Separation and Purification Technology*. 2007, 58, 211-218.

[51] Deng, W; Carpenter, C; Yi, N; Flytzani-Stephanopoulos, M. Comparison of the activity of Au/CeO_2 and Au/Fe_2O_3 catalysts for the CO oxidation and the water-gas shift reactions. *Topics in Catalysis*. 2007, 44, 199-208.

[52] Aguilar-Guerrero, V; Gates, BC. Genesis of a highly active cerium oxide-supported gold catalyst for CO oxidation. *Chemical Communications*. 2007, 3210-3212.

[53] Ko, E-Y; Park, ED; Lee, HC; Lee, D; Kim, S. Nano-sized Au/CeO_2 catalysts for total and selective CO oxidation. In: Ahn, BT, Jeon, H, Hur, BY, Kim, K, Park, JW, editors. *Advances in Nanomaterials and Processing*, Pts 1 and 2. vol. 124-1262007. p. 1749-1752.

[54] Romero-Sarria, F; Martinez, LM; Centeno, MA; Odriozola, JA. Surface dynamics of Au/CeO_2 catalysts during CO oxidation. *Journal of Physical Chemistry* C. 2007, 111, 14469-14475.

[55] Shapovalov, V; Metiu, H. Catalysis by doped oxides: CO oxidation by $Au_xCe_{1-x}O_2$. *Journal of Catalysis*. 2007, 245, 205-214.

[56] Sun, C; Li, H; Chen, L. Study of flowerlike CeO2 microspheres used as catalyst supports for CO oxidation reaction. *Journal of Physics and Chemistry of Solids*. 2007, 68, 1785-1790.

[57] Widmann, D; Leppelt, R; Behm, RJ. Activation of a Au/CeO_2 catalyst for the CO oxidation reaction by surface oxygen removal/oxygen vacancy formation. *Journal of Catalysis*. 2007, 251, 437-442.

[58] Aguilar-Guerrero, V; Gates, BC. Kinetics of CO oxidation catalyzed by highly dispersed CeO_2-supported gold. *Journal of Catalysis*. 2008, 260, 351-357.

[59] Chen, Z; Gao, Q. Enhanced carbon monoxide oxidation activity over gold-ceria nanocomposites. *Applied Catalysis B-Environmental*. 2008, 84, 790-796.

[60] Jia, K; Zhang, H; Li, W. Effect of the Morphology of the Ceria Support on the Activity of Au/CeO_2 Catalysts for CO Oxidation. *Chinese Journal of Catalysis*. 2008, 29, 1089-1092.

[61] Zhou, Z; Kooi, S; Flytzani-Stephanopoulos, M; Saltsburg, H. The Role of the Interface in CO Oxidation on Au/CeO_2 Multilayer Nanotowers. *Advanced Functional Materials*. 2008, 18, 2801-2807.

[62] Huang, X-S; Sun, H; Wang, L-C; Liu, Y-M; Fan, K-N; Cao, Y. Morphology effects of nanoscale ceria on the activity of Au/CeO_2 catalysts for low-temperature CO oxidation. *Applied Catalysis B-Environmental*. 2009, 90, 224-232.

[63] Niu, F; Zhang, D; Shi, L; He, X; Li, H; Mai, H; Yan, T. Facile synthesis, characterization and low-temperature catalytic performance of Au/CeO_2 nanorods. *Materials Letters*. 2009, 63, 2132-2135.

[64] Zhang, H; Yan, X; Li, W. Nanocast Ordered Mesoporous CeO_2 as Support for Highly Active Gold Catalyst in CO Oxidation. *Chinese Journal of Catalysis*. 2009, 30, 1085-1090.

[65] Han, M; Wang, X; Shen, Y; Tang, C; Li, G; Smith, RL, Jr. Preparation of Highly Active, Low Au-Loaded, Au/CeO_2 Nanoparticle Catalysts That Promote CO Oxidation at Ambient Temperatures. *Journal of Physical Chemistry* C. 2010, 114, 793-798.

[66] Sun, Y-a; Shen, Y-n; Jia, M-l; Guo, J-l. Evolution of Gold Species in an Au/CeO_2 Catalyst and Its Impact on Activity for CO Oxidation. *Chemical Research in Chinese Universities*. 2010, 26, 453-459.

[67] Guan, Y; Ligthart, DAJM; Pirgon-Galin, O; Pieterse, JAZ; van Santen, RA; Hensen, EJM. Gold stabilized by nanostructured ceria supports: Nature of the active sites and catalytic performance. *Topics in Catalysis*. 2011, 54, 424-438.

[68] Wang, L; Guo, G; Gu, F; Wang, Z. Preparation of three different sphere-like Au/CeO_2 catalysts and their activity for the CO oxidation. In: Zhang, GJXJ, editor. *Materials Science and Engineering Applications*, Pts 1-3. vol. 160-1622011. p. 428-433.

[69] Gardner, SD; Hoflund, GB; Davidson, MR; Laitinen, HA; Schryer, DR; Upchurch, BT. Catalytic behavior of noble-metal reducible oxide materials for low-temperature CO oxidation. 2. Surface characterization of Au/MnO_x. *Langmuir*. 1991, 7, 2140-2145.

[70] Gardner, SD; Hoflund, GB; Schryer, DR; Schryer, J; Upchurch, BT; Kielin, EJ. Catalytic behavior of noble-metal reducible oxide materials for low-temperature CO oxidation.1. Comparison of catalysts performance. *Langmuir*. 1991, 7, 2135-2139.

[71] Hoflund, GB; Gardner, SD; Schryer, DR; Upchurch, BT; Kielin, EJ. Au/MnO_x catalytic performance - Characteristics for low-temperature carbon-monoxide oxidation. *Applied Catalysis B-Environmental*. 1995, 6, 117-126.

[72] Hoflund, GB; Gardner, SD; Schryer, DR; Upchurch, BT; Kielin, EJ. Influence of promoters on the performance of Au/MnO_x and $Pt/SnO_x/SiO_2$ low-temperature CO oxidation catalysts. *Reaction Kinetics and Catalysis Letters*. 1996, 58, 19-26.

[73] Lee, SJ; Gavriilidis, A; Pankhurst, QA; Kyek, A; Wagner, FE; Wong, PCL; Yeung, KL. Effect of drying conditions of Au-Mn co-precipitates for low-temperature CO oxidation. *Journal of Catalysis*. 2001, 200, 298-308.

[74] Abecassis-Wolfovich, M; Jothiramalingam, R; Landau, MV; Herskowitz, M; Viswanathan, B; Varadarajan, TK. Cerium incorporated ordered manganese oxide OMS-2 materials: Improved catalysts for wet oxidation of phenol compounds. *Applied Catalysis B-Environmental*. 2005, 59, 91-98.

[75] Sinha, AK; Suzuki, K; Takahara, M; Azuma, H; Nonaka, T; Fukumoto, K. Mesostructured manganese oxide/gold nanoparticle composites for extensive air purification. *Angewandte Chemie-International Edition*. 2007, 46, 2891-2894.

[76] Wang, L-C; Huang, X-S; Liu, Q; Liu, Y-M; Cao, Y; He, H-Y; Fan, K-N; Zhuang, J-H. Gold nanoparticles deposited on manganese(III) oxide as novel efficient catalyst for low temperature CO oxidation. *Journal of Catalysis*. 2008, 259, 66-74.

[77] Wang, L-C; Liu, Q; Huang, X-S; Liu, Y-M; Cao, Y; Fan, K-N. Gold nanoparticles supported on manganese oxides for low-temperature CO oxidation. *Applied Catalysis B-Environmental*. 2009, 88, 204-212.

[78] Longo, A; Liotta, LF; Di Carlo, G; Giannici, F; Venezia, AM; Martorana, A. Structure and the Metal Support Interaction of the Au/Mn Oxide Catalysts. *Chemistry of Materials*. 2010, 22, 3952-3960.

[79] Kikegawa, S; Matsumura, T; Murayama, S; Takano, H; Obi, Y. Magnetic properties of amorphous $Mn_{100-x}Ce_x$ alloys. *Physica B-Condensed Matter*. 2003, 329, 1074-1076.

[80] Kikegawa, S; Amakai, Y; Asano, K; Murayama, S; Takano, H; Obi, Y. Specific heat of amorphous $Mn_{100-x}Ce_x$ alloys. *Journal of Magnetism and Magnetic Materials*. 2004, 272, 72-74.

[81] Amakai, Y; Murayama, S; Takanao, H; Okada, Y; Asano, K; Obi, Y. Y substitution effect to the dense Kondo behavior of amorphous Ce-Mn alloys. *Physica B-Condensed Matter*. 2006, 378-80, 854-856.

[82] Ye, Q; Xu, BQ. Textural and structure characterizations of $Ce_{1-x}Mn_xO_2$ prepared by citric acid sol-gel method. *Acta Physico-Chimica Sinica*. 2006, 22, 345-349.

[83] Zhang, H; Yang, W; Li, D; Wang, X. Influence of preparation method on the performance of Mn-Ce-O catalysts. *Reaction Kinetics and Catalysis Letters*. 2009, 97, 263-268.

[84] Imamura, S; Nakamua, M; Kawabata, N; Yoshida, K; Ishida, S. Wet oxidation of poly(ethylene glycol) catalyzed by manganese-cerium composite oxide. *Industrial and Engineering Chemistry Research*. 1986, 25, 34-37.

[85] Imamura, S. Catalytic and Noncatalytic Wet Oxidation. Industrial and *Engineering Chemistry Research*. 1999, 38, 1743-1753.

[86] Chen, H; Sayari, A; Adnot, A; Larachi, F. Composition-activity effects of Mn-Ce-O composites on phenol catalytic wet oxidation. *Applied Catalysis B-Environmental*. 2001, 32, 195-204.

[87] Larachi, F; Pierre, J; Adnot, A; Bernis, A. Ce 3d XPS study of composite $Ce_xMn_{1-x}O_{2-y}$ wet oxidation catalysts. *Applied Surface Science*. 2002, 195, 236-250.

[88] Silva, AMT; Castelo-Branco, IM; Quinta-Ferreira, RM; Levec, J. Catalytic studies in wet oxidation of effluents from formaldehyde industry. *Chemical Engineering Science*. 2003, 58, 963-970.

[89] Silva, AMT; Oliveira, ACM; Quinta-Ferreira, RM. Catalytic wet oxidation of ethylene glycol: kinetics of reaction on a Mn-Ce-O catalyst. *Chemical Engineering Science*. 2004, 59, 5291-5299.

[90] Silva, AMT; Marques, RRN; Quinta-Ferreira, RM. Catalysts based in cerium oxide for wet oxidation of acrylic acid in the prevention of environmental risks. *Applied Catalysis B: Environmental* 2004, 47, 269-279.

[91] Blanco, G; Cauqui, MA; Delgado, JJ; Galtayries, A; Perez-Omil, JA; Rodriguez-Izquierdo, JM. Preparation and characterization of Ce-Mn-O composites with applications in catalytic wet oxidation processes. *Surface and Interface Analysis*. 2004, 36, 752-755.

[92] Abecassis-Wolfovich, M; Landau, MV; Brenner, A; Herskowitz, M. Catalytic wet oxidation of phenol with Mn-Ce-based oxide catalysts: Impact of reactive adsorption on TOC removal. *Industrial and Engineering Chemistry Research*. 2004, 43, 5089-5097.

[93] Abecassis-Wolfovich, M; Landau, MV; Brenner, A; Herskowitz, M. Low-temperature combustion of 2,4,6-trichlorophenol in catalytic wet oxidation with nanocasted Mn-Ce-oxide catalyst. *Journal of Catalysis*. 2007, 247, 201-213.

[94] Arena, F; Trunfio, G; Negro, J; Spadaro, L. Optimization of the $MnCeO_x$ system for the catalytic wet oxidation of phenol with oxygen (CWAO). *Applied Catalysis B-Environmental*. 2008, 85, 40-47.

[95] Levi, R; Milman, M; Landau, MV; Brenner, A; Herskowitz, M. Catalytic wet air oxidation of aniline with nanocasted Mn-Ce-oxide catalyst. *Environmental Science and Technology*. 2008, 42, 5165-5170.

[96] Perez-Omil, JA; Delgado, JJ; Ouahbi, W; Hungria, AB; Browning, N; Cauqui, MA; Rodriguez-Izquierdo, JM; Calvino, JJ. Electron microscopy investigations of nanostructured Ce/Mn oxides for catalytic wet oxidation. *Journal of Physical Chemistry* C. 2010, 114, 8981-8991.

[97] Ye, Q; Gao, Q; Zhang, XR; Xu, BQ. Cataluminescent characteristics of catalytic ethanol oxidation over $Ce_{1-x}Mn_xO_2$ catalysts. *Acta Chimica Sinica*. 2006, 64, 751-755.

[98] Rao, T; Shen, M; Jia, L; Hao, J; Wang, J. Oxidation of ethanol over Mn-Ce-O and Mn-Ce-Zr-O complex compounds synthesized by sol-gel method. *Catalysis Communications*. 2007, 8, 1743-1747.

[99] Tang, Q; Wu, C; Huang, X; Yang, Y. Selective Oxidation of Aromatic Alcohols over Amorphous Mn-Ce-O Catalyst. *Chinese Journal of Catalysis*. 2009, 30, 207-212.

[100] Chen, M; Shi, C-M; Zheng, X-M. Properties and coating effect of nano-powder of Ce-Mn-O for VOCs catalysts. *Chinese Journal of Inorganic Chemistry*. 2006, 22, 1828-1832.

[101] Delimaris, D; Ioannides, T. VOC oxidation over MnO_x-CeO_2 catalysts prepared by a combustion method. *Applied Catalysis B-Environmental*. 2008, 84, 303-312.

[102] Deng, S; Liu, H; Zhou, W; Huang, J; Yu, G. Mn-Ce oxide as a high-capacity adsorbent for fluoride removal from water. *Journal of Hazardous Materials*. 2011, 186, 1360-1366.

[103] Qi, GS; Yang, RT. Performance and kinetics study for low-temperature SCR of NO with NH_3 over MnO_x-CeO_2 catalyst. *Journal of Catalysis*. 2003, 217, 434-441.

[104] Qi, GS; Yang, RT; Chang, R. MnO_x-CeO_2 mixed oxides prepared by co-precipitation for selective catalytic reduction of NO with NH_3 at low temperatures. *Applied Catalysis B-Environmental.* 2004, 51, 93-106.

[105] Eigenmann, F; Maciejewski, M; Baiker, A. Selective reduction of NO by NH_3 over manganese-cerium mixed oxides: Relation between adsorption, redox and catalytic behavior. *Applied Catalysis B-Environmental.* 2006, 62, 311-318.

[106] Tang, X; Hao, J; Yi, H; Ning, P. Low-temperature SCR of NO with NH_3 on Mn-based catalysts modified with cerium. *Journal of Rare Earths.* 2007, 25, 240-243.

[107] Li, H; Tang, X; Yi, H; Yu, L. Low-temperature catalytic oxidation of NO over Mn-Ce-O_x catalyst. *Journal of Rare Earths.* 2010, 28, 64-68.

[108] Martins, RCC; Leal, HM; Quinta-Ferreira, RMO; Iaeng. *Catalytic removal of phenolic compounds by ozone using manganese and cerium oxides.* Ao, SI, Douglas, C, Grundfest, WS, Schruben, L, Wu, X, editors. vol. 2007.

[109] Martins, RC; Quinta-Ferreira, RM. Catalytic ozonation of phenolic acids over a Mn-Ce-O catalyst. *Applied Catalysis B-Environmental.* 2009, 90, 268-277.

[110] Picasso, G; Gutierrez, A; Pina, MP; Herguido, J. Preparation and charactrization of Ce-Zr and Ce-Mn based oxides for n-hexane combustion: Application to catalytic membrane reactors. *Chemical Engineering Journal.* 2007, 126, 119-130.

[111] Shi, L; Chu, W; Qu, F; Hu, J; Li, M. Catalytic performance for methane combustion of supported Mn-Ce mixed oxides. Journal of Rare Earths. 2008, 26, 836-840.

[112] Wang, X; Kang, Q; Li, D. Low-temperature catalytic combustion of chlorobenzene over MnO_x-CeO_2 mixed oxide catalysts. *Catalysis Communications.* 2008, 9, 2158-2162.

[113] Zhaing, D-J; Murata, Y; Kishikawa, K; Ikeue, K; Machida, M. Synthesis of large surface area MnO_x-CeO_2 using CTA and its catalytic activity for soot combustion. *Journal of the Ceramic Society of Japan.* 2008, 116, 230-233.

[114] Wang, X; Kang, Q; Li, D. Catalytic combustion of chlorobenzene over MnO_x-CeO_2 mixed oxide catalysts. *Applied Catalysis B-Environmental.* 2009, 86, 166-175.

[115] Wu, Y; Zhang, Y; Liu, M; Ma, Z. Complete catalytic oxidation of o-xylene over Mn-Ce oxides prepared using a redox-precipitation method. *Catalysis Today.* 2010, 153, 170-175.

[116] Guo, Q; Liu, Y. MnOx modified Co_3O_4-CeO_2 catalysts for the preferential oxidation of CO in H_2-rich gases. *Applied Catalysis B-Environmental.* 2008, 82, 19-26.

[117] Guo, Q; Tang, Z; Liu, Y. Low-temperature CO Oxidation over Coprecipitated Co-Ce-Mn Oxides. *Chemistry Letters.* 2008, 37, 1274-1275.

[118] Guo, Q; Chen, S; Liu, Y; Wang, Y. Stability of Co-Ce-Mn mixed-oxide catalysts for CO preferential oxidation in H_2-rich gases. *Chemical Engineering Journal.* 2010, 165, 846-850.

[119] Chang, LH; Sasirekha, N; Chen, YW; Wang, WJ. Preferential oxidation of CO in H_2 stream over Au/MnO_2-CeO_2 catalysts. *Industrial and Engineering Chemistry Research.* 2006, 45, 4927-4935.

[120] Tu, Y-B; Luo, J-Y; Meng, M; Wang, G; He, J-J. Ultrasonic-assisted synthesis of highly active catalyst Au/MnO_x-CeO_2 used for the preferential oxidation of CO in H_2-rich stream. *International Journal of Hydrogen Energy.* 2009, 34, 3743-3754.

[121] Meng, M; Tu, Y; Ding, T; Sun, Z; Zhang, L. Effect of synthesis pH and Au loading on the CO preferential oxidation performance of Au/MnO_x-CeO_2 catalysts prepared with ultrasonic assistance. *International Journal of Hydrogen Energy.* 2011, 36, 9139-9150.

[122] Bowker, M; Nuhu, A; Soares, J. High activity supported gold catalysts by incipient wetness impregnation. *Catalysis Today.* 2007, 122, 245-247.

In: Carbon Monoxide: Sources, Uses and Hazards ISBN:978-1-61942-055-7
Editors: D. DiLoreto et al.

Chapter 4

HETEROGENEOUS CATALYSIS IN CARBONYLATIVE COUPLING REACTIONS

Mayur V. Khedkar, Shoeb R. Khan and Bhalchandra M. Bhanage*

Department of Chemistry, Institute of Chemical Technology, Mumbai, India

ABSTRACT

Carbon monoxide (CO) in combination with transition metal catalysts has become a versatile reagent in organic synthesis. Insertion of the CO molecule introduces a carbonyl group, which is one of the most common functionalities in bioactive/industrially important compounds. The bulk chemical industry also takes advantage of CO as one of the cost efficient C_1 building block for various transformations like olefins into aldehydes, carboxylic acid derivatives, or alcohols by way of carbonylation reactions, thus providing a straight forward access to valuable intermediates for the synthesis of polymers and other consumables.

After the first reports by Heck and co-workers in the 1970s on palladium catalyzed carbonylation reactions, the scope of CO chemistry has expanded considerably. Since then various new classes of carbonylation reactions such as alkoxycarbonylation,

* Tel.: +91 22 24145616; fax: +91 22 24145614; Email address: bm.bhanage@gmail.com, bm.bhanage@ictmumbai.edu.in

phenoxycarbonylation, aminocarbonylation, thiocarbonylation and carbonylative coupling reactions like carbonylative Suzuki coupling, carbonylative Sonogashira coupling, carbonylative Heck coupling etc. have been emerged and mainly using palladium as a catalyst of choice with good catalytic activity and selectivity. Several research groups have developed various homogeneous palladium complexes based catalytic systems for these carbonylation reactions in which catalyst are in dissolved with the substrate. Despite of their potential utility, homogeneous catalytic systems have one or more drawback such as difficulties in catalyst and product separation, loss of expensive metal and non-recyclability of metal catalyst limiting their applications. Therefore, to craft synthetic protocol more economical and efficient, heterogeneous catalysis was found to be the best alternative to homogeneous processes. Also the catalyst recovery decreases the contamination of desired products with trace metal or ligand impurities. The contamination of catalyst even in ppm concentration is not desirable for pharmaceutical applications. Herein, we summarize the recent developments in use of heterogeneous catalysts for these new classes of carbonylative coupling reactions.

INTRODUCTION

Heterogeneous catalysis influenced our lives greatly in 20th century. In 1908, the German chemist Fritz Harber firstly introduces application of Osmium catalyst in synthesizing ammonia by feeding N_2 and H_2 at high pressure. The ammonia synthesis using Osmium catalyst has became one of the most important chemical processes worldwide, earning Harber the Nobel Prize in 1918 in the field of chemistry. Since then catalysis has played an important role in the development of chemical industry, which includes developing new technologies, production of valuable chemicals in economical ways and solving environmental problems. Predominantly, the advances in the conversion of cheaper feedstock to value-added chemical products, new synthetic routes for pharmaceuticals, fine chemicals and specialty chemicals to replace stoichiometric routes, waste minimization, pollutants removal, global preservation and supply of energy have made significant impact [1].

Catalysis has undergone significant development from fundamental knowledge and of its applications point of view. Catalysis is the heart of chemical sciences because it is the art of selectively making and breaking of chemical bonds. The design and applications of new catalysts and catalytic systems are simultaneously achieving the dual goals of environmental protection and economic benefit. Catalysis is also a fundamental component of the innovation towards cleaner production processes. Catalysis can not only help to make the chemical processes greener but also reduces the environmental impact and the cost of these processes.

Catalysts are broadly classified into to three main categories depending upon the nature in which they are present in the reaction system.

a) Homogeneous catalyst
b) Heterogeneous catalyst
c) Heterogenised homogeneous catalyst

(a) Homogeneous Catalyst

Homogeneous catalyst is defined as one in which all the constituents of the reaction are present in the same phase including the catalyst. Homogenous catalysis has been used efficiently for years. Various types of transition metal salts and complexes ($TiCl_4$, $Bi(NO_3)_3$, $Pd(PPh_3)_4$, $PdCl_2(PPh_3)_2$, CuI/DABCO, $Cu(OAc)_2$ / 1,10-phenanthroline, alkali and alkaline earth metal halides ($AlCl_3$, $MgCl_2$), metal hydroxides, phase transfer catalyst, ionic liquids etc. are most commonly employed homogenous catalyst in organic transformations.

Advantages of homogeneous catalyst are as follows:

- More effective and specific
- High activity and selectivity
- Availability of all active sites
- Ease of monitoring of the reaction

The major disadvantage of homogeneous catalysis is the difficulty in separation of catalyst from the reaction mixture and product. The separation and purification of the products from reaction mixture are time consuming and tedious. Many times homogeneous catalyst are moisture sensitive, thermally unstable, corrosive and create handling problems. Homogeneous catalytic systems generally have major concern with recycling of precise metal catalyst.

(b) Heterogeneous Catalyst

Heterogeneous catalyst is the one in which one or more of the constituents are in different phases. The heterogeneous catalysts are classified into supported metal catalysts, zeolite-encapsulated catalysts, colloids–nanoparticles, and intercalated metal compounds. In heterogeneous catalysis the catalyst is usually present as a solid and the reactants are liquid or more frequently gases.

Advantages of heterogeneous catalysts are as follows:

- Ease of handling.
- Catalyst recyclability.
- High thermal stability.
- Possibility of carrying out the reaction in continuous mode.

Various mesoporous materials (MCM-41, MCM-42, zeolites, clays), polymeric materials (poly vinyl pyridine), heteropoly acids, ion exchange resin (Amberlyst-15, Nafion NR-50), metal oxides, hydrotalcites etc. are most commonly employed heterogeneous catalysts in organic transformations.

The disadvantages of heterogeneous catalysts are low concentration of the surface sites, difficulty in reproducing the results, difficulty in agitation of reaction mixture and diffusion barrier for the reactants on the catalysts surface.

(c) Heterogenised Homogeneous Catalyst

Homogeneous transition metal complex catalysts are superior in activity, selectivity as well as specificity at mild reaction conditions [2]. However; a clear drawback with the use of homogeneous catalysts is the difficulty of separating them from the product mixture.

This step often relies on classical separation techniques such as distillation or adsorption, which can result in the degradation or total destruction of the catalyst. Hence it would be of great environmental, commercial and scientific interest to combine the advantages of homogeneous and heterogeneous catalysts. Consequently more efforts have been taken to heterogenise the homogeneous catalysts.

To counteract these drawbacks, catalysts are being developed that are specifically designed to facilitate their separation while leaving the catalytic properties unharmed. Homogeneous metal catalysts are heterogenised using different techniques are known as heterogenised homogeneous catalyst for example; modified silica catalysts, polymer-supported catalysts, biphasic catalysts, supported liquid-phase catalysts, non ionic liquids solvents, perfluorinated solvents, and reusable homogeneous complexes.

CARBONYLATION

The term carbonylation was coined by W. Reppe during the thirties and is generally used to cover a large number of closely related reactions that all have in common the carbon monoxide which is incorporated into a substrate like an aryl, benzyl or vinyl palladium complex in presence of various nucleophiles. CO is a high affinity ligand for palladium, in oxidation states 0 and II, by its dual ability to act as an σ-donor and π-acceptor. Carbonylation reactions rank among the most useful transformations catalyzed by transition metal complexes, forming the basis for industrial and laboratory processes currently in practice [3-6]. The classical examples of Pd-catalyzed carbonylative coupling reactions using halides or pseudo-halides with a suitable nucleophile include alkoxycarbonylation, aminocarbonylation and various carbonylative coupling reactions such as carbonylative Sonogashira coupling reactions, carbonylative Suzuki-Miyaura, carbonylative stille, carbonylative Heck coupling reactions and so forth (figure 1).

These transition metal catalyzed carbonylation reactions hold the advantage of introducing the carbonyl functionality in one step by electrophilic means. Traditionally, carbonyl functionalities are installed by trapping carbon dioxide with a suitable nucleophile, being typically organolithium, organozinc, or Grignard reagents, and subsequently transforming the formed acid into the desired product. The latter method compromises the functional group tolerance of the system due to the presence of highly basic or nucleophilic reagents and requires the carbonyl moiety to be installed early in the synthesis.

By use of palladium catalysis, the acid derivatives can be installed almost at any desired step in a synthesis due to the mild and selective conditions under which carbonylations are performed. The carboxylic acid and its derivatives like amides, esters, thioamides and ketones are prepared via carbonylation reactions and they are important intermediates in the manufacture of dyes, pharmaceuticals, agrochemicals, and other industrial products. Recently transition metal catalysed carbonylative C-H activation reactions are also explored.

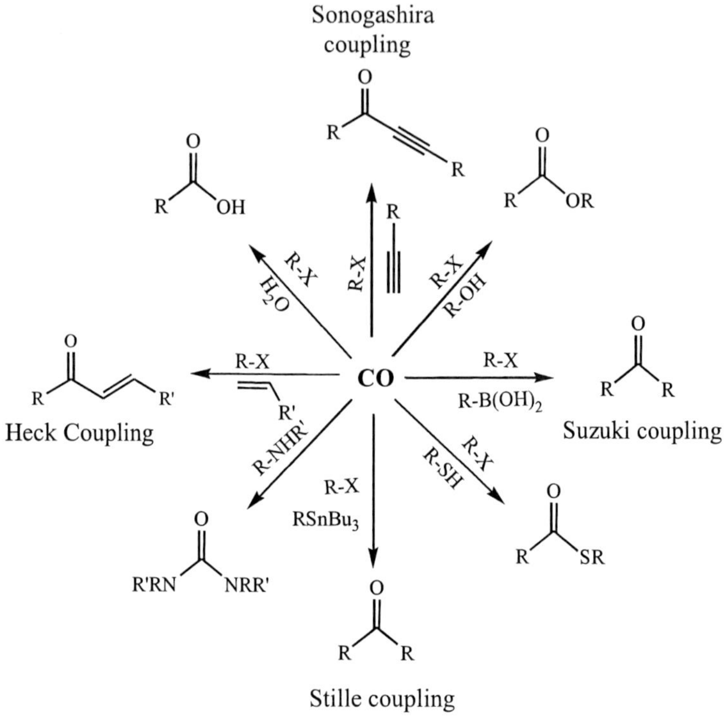

Figure 1. Transition metal catalyzed carbonylation reactions.

Carbon-carbon bond cross-coupling reactions are among the most useful and most widely studied synthetic transformations. Most important ones are carbonylative Sonogashira reaction, carbonylative Suzuki-Miyaura coupling, carbonylative Stille reaction and carbonylative Heck reaction. They are characteristically catalyzed by palladium and abundantly used in synthesis. The importance of palladium catalyzed C-C bond coupling reactions has been recognized by awarding the Nobel Prize in 2010 to Professors Heck, Negishi, and Suzuki in the field of chemistry. Some of important contributions using heterogeneous catalysis in carbonylative coupling reactions are summarized below.

Carbonylative Sonogashira Coupling Reaction

α,β-alkynyl ketones are important intermediates in organic syntheses, and due to their bifunctional electrophilicity, they found broad application as three-carbon building blocks in heterocyclic synthesis. They are crucial moieties in many biologically active molecules, natural products and pharmaceuticals [7]. Traditionally α,β-alkynyl ketones were synthesized via the coupling of alkynyl organometallic reagents with acid chlorides [8]. An atom economic route to prepare these compounds is carbonylative Sonogashira coupling reaction: metal-catalyzed direct coupling of aryl halides with terminal alkynes in the presence of carbon monoxide. Some heterogeneous and recyclable protocols developed for carbonylative Sonogashira coupling reaction are discussed below.

Pd/C is an inexpensive and readily available Pd metal source; it was explored as a heterogeneous, ligand free and reusable catalyst for carbonylative Sonogashira coupling of terminal alkynes with aryl iodides using triethyl amine as a base in MeOH (Scheme 1) by Jianhua Liu and co-workers [9].

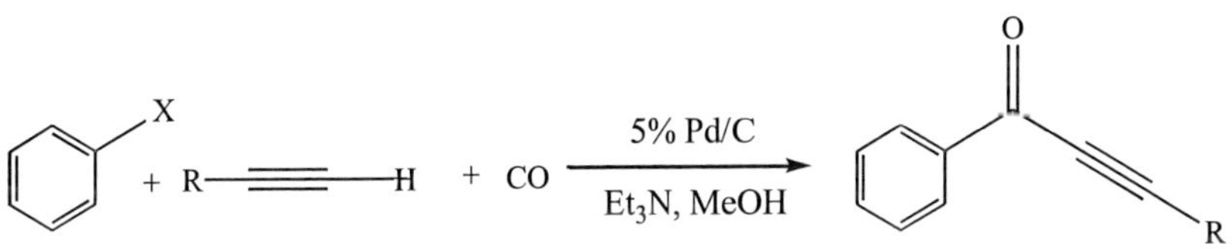

Scheme 1. Palladium catalyzed carbonylative Sonogashira reactions.

Chungu Xia et al. reported magnetically separable Pd-Fe_3O_4 catalyst (Scheme 2) [10]. This superparamagnetic Pd-Fe_3O_4 catalyst, prepared by wet impregnation of Fe_3O_4 with $Na_2[PdCl_4]$ followed by reduction with KBH_4. The catalyst was completely magnetically recoverable due to the super paramagnetic behaviour of Fe_3O_4 and was recycled for seven consecutive cycles after magnetic separation.

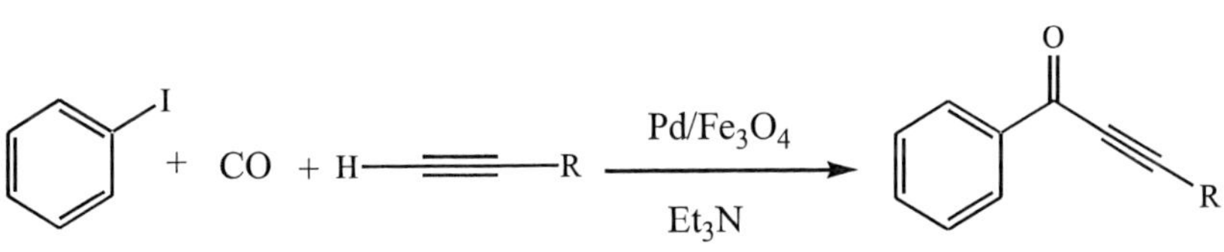

Scheme 2. Carbonylative Sonogashira reactions using Pd-Fe_3O_4 catalyst.

Marie Genelot and co-workers developed one-pot tandem procedure for the selective synthesis of 2-benzylidene-indoxyl and 2-phenyl-4-quinolone by intramolecular Sonogashira coupling (Scheme 3). They developed heterogeneous palladium-, amine-, and phosphine-catalysts by post-synthetic grafting onto SBA silica [11].

The catalyst exhibit high activities and, 2-benzylidene-indoxyl was obtained in 81% yields through {[Pd]@SBA-15+PPh_3} catalysis while 2-phenyl-4-quinolone was prepared by a fully heterogeneous {[Pd]@SBA-15+[AMINE]@SBA-3} catalyst in 61–75% isolated yield. Catalyst was further recycled for three consecutive cycles.

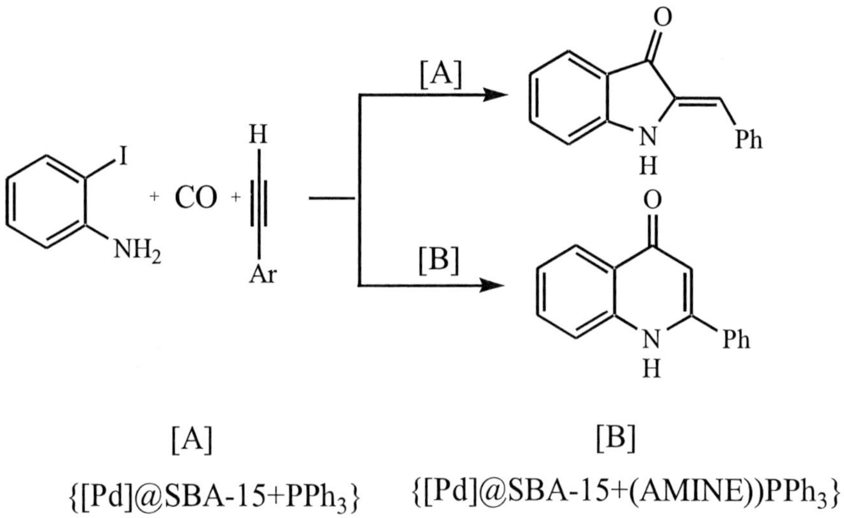

Scheme 3. Intramolecular carbonylative Sonogashira coupling reactions.

Carbonylative Sonogashira coupling reaction of aryl iodides with alkynes for the synthesis of alkynyl ketones by using water as a solvent was investigated by Zhen Yang and co-workers (Scheme 4) [12]. They extended protocol for Flavones synthesis through carbonylative cyclization but under homogeneous reaction condition.

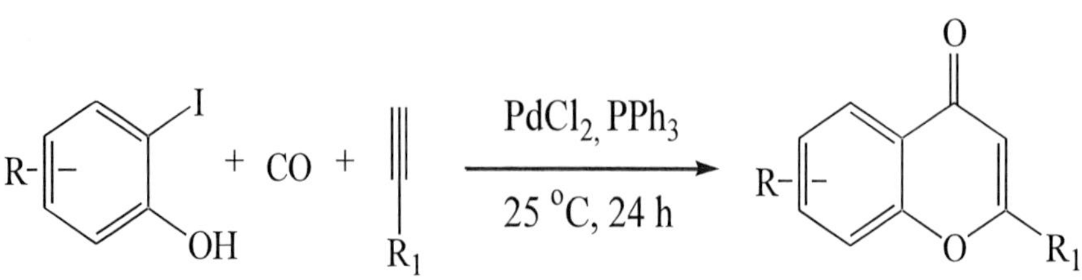

Scheme 4. Palladium catalyzed carbonylative cyclization.

Chungu Xia and co-workers, reported the synthesis of a cross-linked co-polymer P(DVB-IL) supported palladium catalyst and illustrate its application in the carbonylative Sonogashira coupling reaction of aryl iodides with terminal alkynes to form α,β-alkynyl ketones in water (Scheme 5) [13]. P(DVB-IL) and P(DVB-IL)-Pd catalyst were characterized by thermogravimetric analysis, also the catalyst reusability was tested for five consecutive cycles.

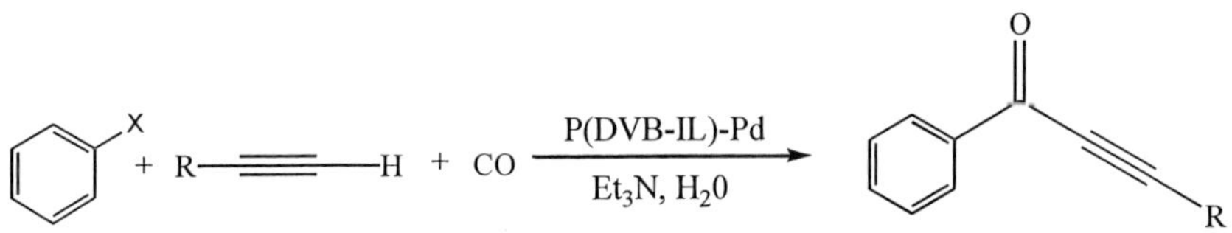

Scheme 5. P(DVB-IL)-Pd catalyzed carbonylative Sonogashira coupling reaction.

Low pressure Pd-catalyzed carbonylation in an ionic liquid using a multiphase microflow system was demonstrated by Ilhyong Ryu et al. (Figure 2) [14]. $[BMIm]PF_6$ ionic liquid was used as reaction media and it was successfully recycled.

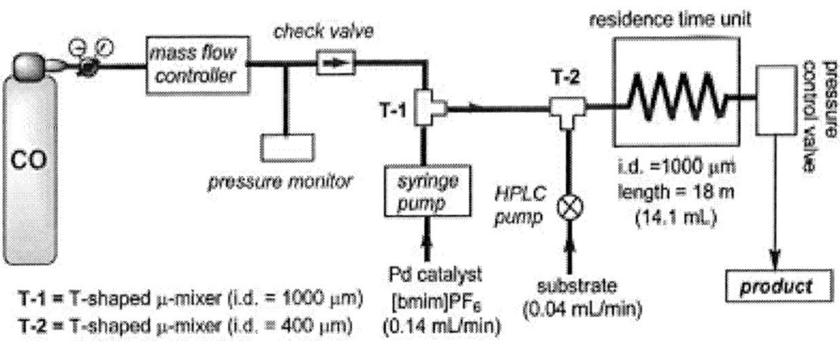

Figure 2. Schematic diagram of the microflow system for carbonylative Sonogashira coupling.

Mobil composition of matter (MCM) is the initial name given for a series of mesoporous materials that were first synthesized by Mobil's researchers in 1992. MCM-41 (Mobil composition of matter No. 41) and MCM-48 (Mobil composition of matter No. 48) are two of the most popular mesoporous molecular sieves that are keenly studied by researchers. The most remarkable fact about the MCM-41 and MCM-48 is that, although composed of amorphous silica wall, they possess long range ordered framework with uniform mesopores. These materials also possess large surface area, which can be up to more than 1000 m^2g^{-1}. Furthermore, the pore diameter of these materials can be nicely controlled within mesoporous range between 1.5 to 20 nm by adjusting the synthesis conditions and/or by employing surfactants with different chain lengths in their preparation. To date, a few palladium complexes on functionalized MCM-41 support have been prepared and successfully used in organic reactions (Figure 3) [15-19].

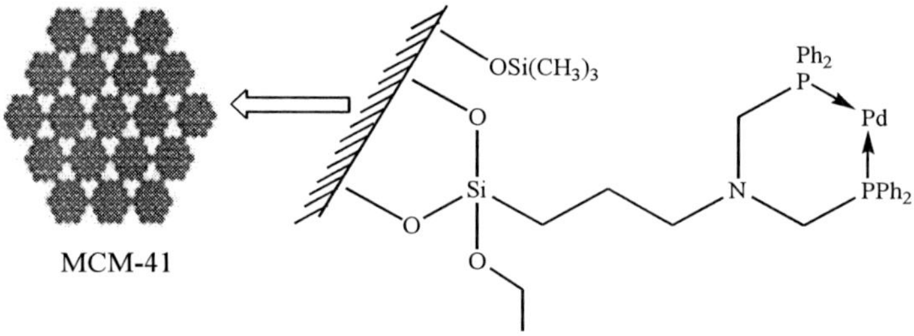

Figure 3. MCM-41-Pd complex.

Recently MCM-41 has found a new possible candidate for a solid support for the immobilization of homogeneous catalysts.

Hao et al. showed application of MCM-41-supported bidentate phosphine palladium(0) complex (Figure 3) for the synthesis of α,β-alkynyl ketones (Scheme 6) [20]. They have developed efficient recyclable catalytic system which can be recycled up to 10 consecutive cycles.

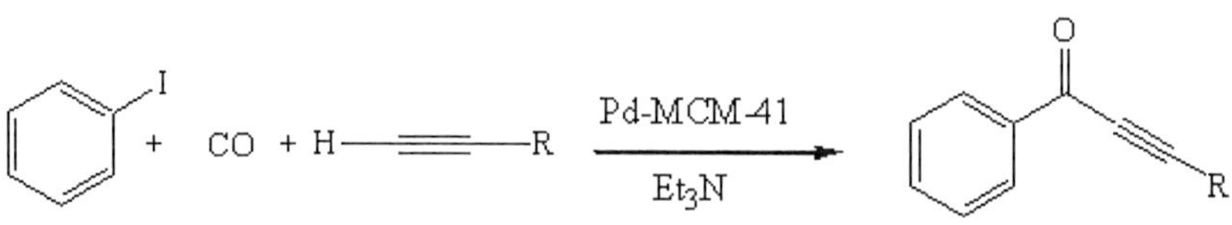

Scheme 6. Pd-MCM41 catalysed carbonylative Sonogashira coupling reaction.

Carbonylative Suzuki Coupling Reaction

The cross-coupling reaction is one of the most straight forward and general methods for the formation of carbon-carbon bonds. In this area, the transition metal catalyzed, three-component, cross-coupling reaction between organometallic reagents, carbon monoxide, and organic halides is now considered as a useful tool for the synthesis of ketones. Biaryl ketones are important moieties in many biologically active molecules (e.g. Suprofen, Ketoprofen), natural products (e.g. Cotoin, Papaveraldine), and pharmaceuticals (Figure 4).

A common route to synthesize biarylketones involves Friedel–Crafts acylation of substituted aromatic ring compounds with acyl halides, which is incompatible with many functional groups and requires over stoichiometric amounts of Lewis acid; furthermore, the regioselectivity is limited to the para position [21].

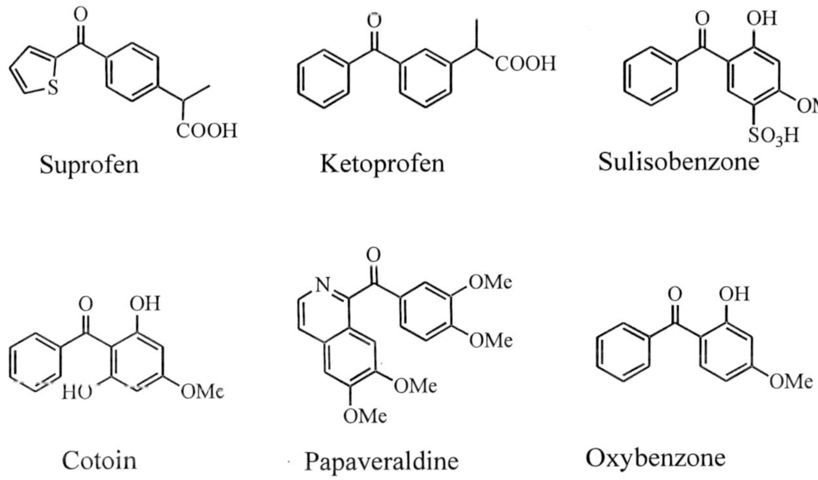

Figure 4. Structure of some important biarylketones derivatives.

Traditionally, biaryl ketones were synthesized by transition metal catalyzed three-component, carbonylative cross coupling reaction between organometallic reagents, carbon monoxide and aryl electrophiles, which is now considered as a useful tool for synthesis of wide variety of carbonyl compounds. Various aryl metal reagents like nickel [22], silicon [23], aluminium [24] and tin [25] have been reported. However, these carbonylative cross-coupling reactions have several limitations due to formation of biaryl side products without carbon monoxide insertion particularly with electron-deficient aryl halides. Other synthetic strategies use cross-coupling reactions of benzoic halides with organotin compounds [26-28].

In 1993, Suzuki et al. reported facial protocol for the synthesis of diaryl ketones from carbon monoxide, aryl halide and aryl boronic acid by using palladium catalyst (Scheme 7) [29]. In principle, this reaction provides a versatile tool for diarylketones synthesis using boronic acids which are generally nontoxic and thermally, air and moisture stable. They used homogeneous $PdCl_2(PPh_3)_2$ as a catalysts and K_2CO_3 as a base for this transformation to give good yield of carbonylated products.

$$\text{Ar-B(OH)}_2 + \text{CO} + \text{I-Ar}' \xrightarrow[\text{Base}]{\text{PdCl}_2(\text{PPh}_3)_2} \text{Ar-CO-Ar}'$$

Scheme 7. Palladium catalyzed Suzuki carbonylation reaction.

After the first report of Suzuki, several groups reported various homogeneous palladium based catalytic systems for the synthesis of such biaryl carbonyl compounds, using $PdCl_2(PPh_3)_2$, $PdCl_2(PPh_3)_2/PdCl_2(dppf)$, $Pd(OAc)_2$-imidazoliumsalts, $Pd(OAc)_2$/N,N-bis(2,6diisopropylphenyl) dihydroimidazoliumchloride, Pd/thiourea and $Pd(OAc)_2$/di-1-adamantyl-n-butylphosphane as a catalyst.

Bhanage and co-workers developed Pd/C as an efficient, heterogeneous and reusable catalyst for phosphine-free carbonylative Suzuki coupling reaction (Scheme 8) [30]. Both aryl and heteroaryl iodides were screened for reaction and found good to excellent yield of expected product. For the first time they reported formation of sterically hindered hetero aryl ketones from iodoindole and iodoquinoline through carbonylative Suzuki cross-coupling. Excellent catalytic activity, reusability and compatibility with various hindered and functionalized aryl/heteroaryl iodides and arylboronic acids makes it an ideal synthetic protocol for carbonylative Suzuki coupling reactions.

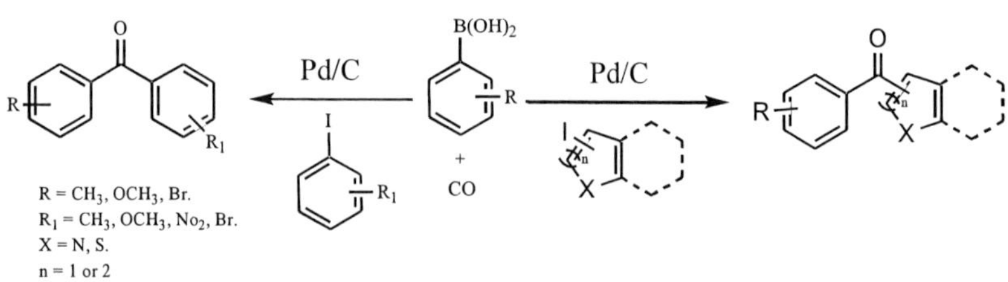

Scheme 8. Pd/C catalyzed Suzuki carbonylation reaction.

Mesoporous material i.e. MCM-41-Pd complex (Figure 3) was used to catalyze the carbonylative coupling of 4-iodoanisole and aryl boronic acids to form aryl ketones by Mingzhong Cai et al. (Scheme 9) [31]. The reaction was optimized with respect to various parameters such as time, temperature, CO pressure and solvent. The catalyst system was applicable for variety of aryl iodides and aryl boronic acids. Furthermore excellent catalyst reusability up to 10 consecutive cycles was shown.

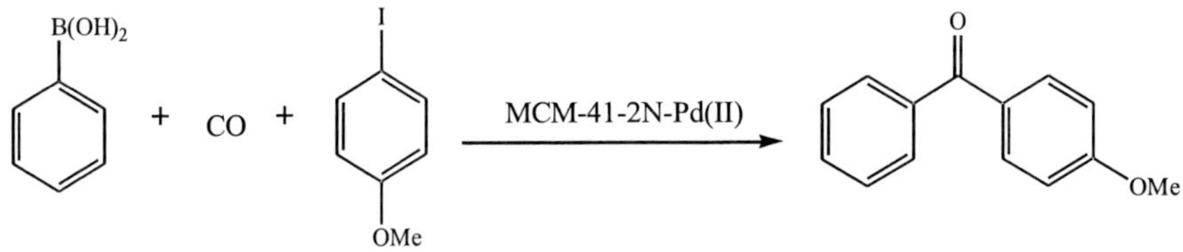

Scheme 9. MCM-41-2N-Pd catalyzed carbonylative Suzuki reaction.

N-Heterocyclic Carbene (NHC) Ligand

Phosphine ligands are well known in Pd-catalyzed cross-coupling reactions. But most of the phosphine ligands are toxic, air/moisture sensitive, and hence synthesis and application of non-toxic and stable *N*-heterocyclic carbenes (NHCs) as a ligand in synthetic chemistry gained much more attention in literature [32]. The concept of using *N*-heterocyclic carbenes (NHCs) as a ligand was first independently introduced by Ofele and Wanzlick. In 1992-93, striking similarity of electron-rich organophosphanes PR_3 and *N*-heterocyclic carbenes (NHCs) in terms of their metal coordination chemistry was noted. It was pointed out that both low and high- oxidation state transition metals are compatible with NHCs. Furthermore, spectroscopic studies also revealed the close relationship of NHCs ligands and organophosphanes.

As compared with many other ligands, NHC ligand show very high dissociation energies, which have been quantified by theoretical calculations for different metals [33]. Therefore, the bonding between the NHC and the transition metal is much stronger and is chemically and thermally more inert toward cleavage as compared to any other complexes. NHC ligands consist of imidazole, pyrazole, or triazole moiety. Various NHC ligand precursors are shown in figure 5.

NHC ligands were well explored in various types of reactions such as C-H activation, C-C, C-H, C-O and C-N bond formation by using it in a homogeneous reaction condition [32]. To take advantage of NHC ligand in heterogeneous catalysis, immobilization of this ligand on various insoluble supports, especially porous materials with high surface areas, is usually the method of choice. Since the immobilized catalysts can be facilely recovered via a simple filtration process after reactions.

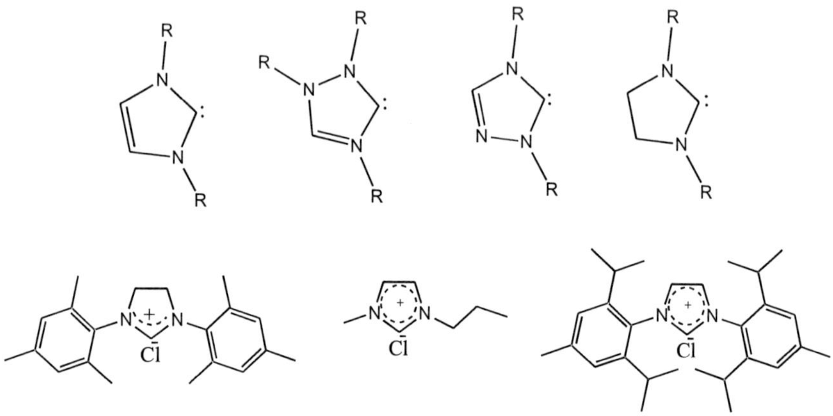

Figure 5. *N*-heterocyclic carbenes precursor.

Due to their versatile processing capabilities, ease of separation and recycling, polymer-supported NHC catalyst offers many advantages for industrial applications. These catalysts can prevent the contamination of the ligand residue in the products. With these concerns, several kinds of polymer-supported NHC were developed recently by anchoring the Pd–NHC complex on to Wang resin [34] or modifying the nitrile groups in PAN fiber into imidazoline groups (Figure 6) [35].

Yong Gao and co-workers further improved catalytic activity of these polymer supported NHC ligand by coating it with superparamagnetic nanoparticle, and its application was reported for Suzuki cross-coupling reactions [36]. To immobilize Pd catalysts, they first loaded 1-methylimidazole on the surface of core/shell nanoparticles, thereafter in second step they prepared Pd metal and magnetically supported ligand complex. Elemental analysis of this formed catalyst confirmed more (0.72 mmol/g) Pd catalysts were loaded on to nanoparticles supports. This is higher than that of Pd on chloromethyl polystyrene resins that were used for immobilization of Pd catalysts by forming similar Pd methylimidazole NHC complexes. In this way NHC ligands were heterogenised and applied for different organic transformations.

However, in many case a considerable decrease in activity and selectivity of the immobilized catalysts is frequently observed due to the heterogeneous nature of the support materials in reaction media. To overcome this problems associated with heterogeneous insoluble matrixes was addressed by using a soluble support like JandaJel [37], perfluorinatedtags [38-39], ROMP gels [40-41], and other types of synthetic soluble polymers [42-43]. In this way NHC ligands were heterogenised and applied in different chemical transformations.

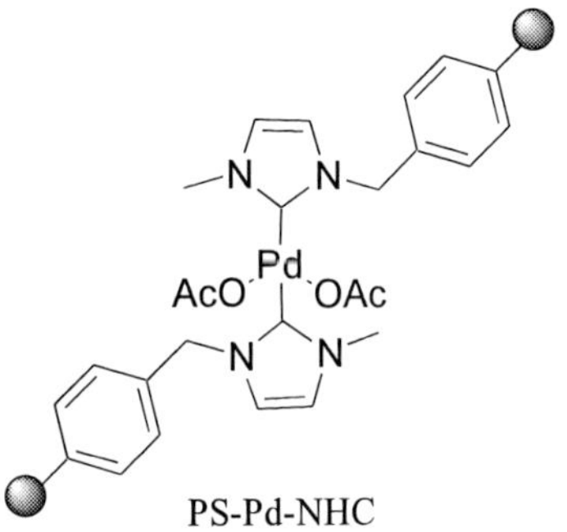

Figure 6. Polymer supported Palladium NHC complex.

Hence some application of NHC ligands in carbonylative coupling reaction are summarised below.

Application of *N*-Heterocyclic Carbene as a Ligand in Carbonylative Coupling Reaction

Shuzhan Zheng et al. described *N*-heterocyclic carbene ligand in palladium catalysed multicomponent carbonylative Suzuki coupling reaction with sodium tetraphenylborate. In this report they described *insitu* generated palladium-*N*-heterocyclic carbene complex (Scheme 10) [44].

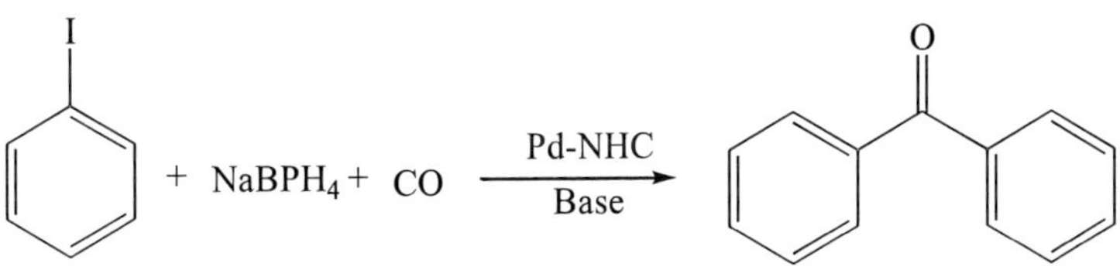

Scheme 10. Pd-NHC catalyzed carbonylative Suzuki reaction.

Andrus and co-workers used Palladium(II) acetate with *N,N*-bis-(2,6-diisopropylphenyl)-dihydroimidazolium chloride i.e. Pd-NHC catalyst (2 mol%) for the carbonylative coupling of aryl diazonium tetrafluoroborate salts and aryl boronic acids to form aryl ketones (Scheme 11) [45]. The reaction was optimized with respect to various parameters such as time, temperature, CO pressure and solvent. The catalyst system was applicable for variety of aryl diazonium tetrafluoroborate salts and aryl boronic acids.

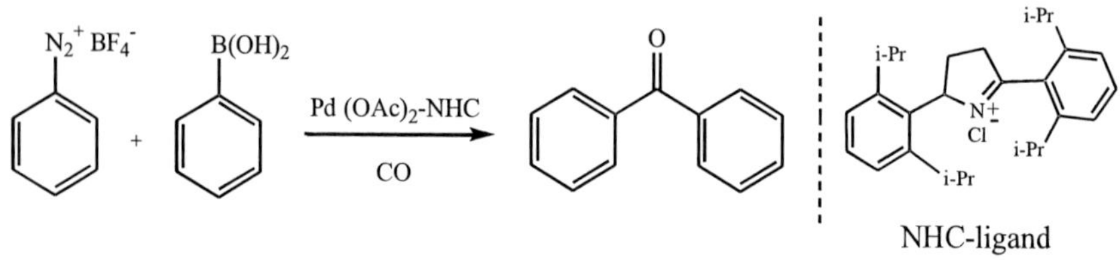

Scheme 11. Carbonylative coupling of aryl diazonium tetrafluoroborate salts and aryl boronic acids using Pd-NHC catalyst.

Castanet et al. reported a simple and efficient method for the synthesis of α-pyridyl ketone from chloropyridines using *insitu* generated Pd-NHC catalyst from $Pd(OAc)_2$/imidazolium salt (Scheme 12) [46]. They have shown applicability of their protocol by coupling chloro- or dichloropyridines and chloroquinolines which practically do not react with other catalytic system based on phosphine with phenylboronic acid to yield the desired product in moderate to good yields (51-86 %).

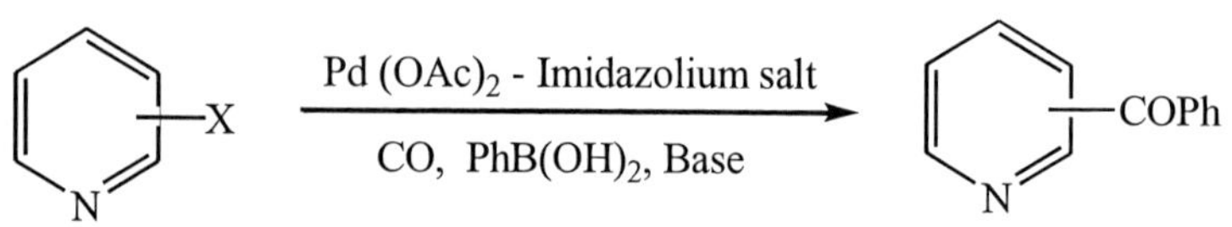

Scheme 12. $Pd(OAc)_2$/imidazolium salt catalyzed carbonylative Suzuki reaction of halopyridines.

Bhanage and co-workers reported the carbonylative Suzuki coupling reaction of aryl and heteroaryl iodides with variety of arylboronic acids catalyzed by this well defined and stable phosphine free palladium complex viz. palladium *N*-heterocyclic carbene as the catalyst (Scheme 13) [47]. The catalyst systems were optimized with respect to various parameters and enabled carbonylation of electron-rich, electron-deficient and sterically hindered aryl iodides, with different types of aryl iodides and phenyl boronic acids affording excellent yields of the desired products. Furthermore catalyst reusability was tested upto four consecutive cycles and also Pd metal leaching was checked by hot filtration and ICP-AES analysis.

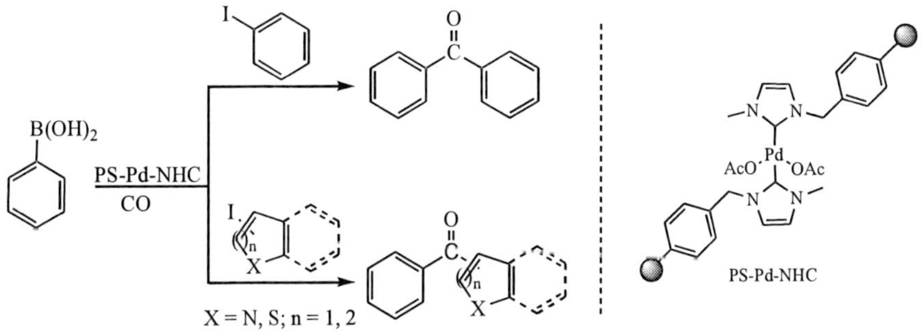

Scheme 13. PS-Pd-NHC catalyzed carbonylative Suzuki reaction.

O'Keffe et al. have recently disclosed a general process for *ortho*-disubstituted aryl ketones employing a commercially available PEPPSI-iPr [a] catalyst, under balloon pressure of CO (Scheme 14) [48].

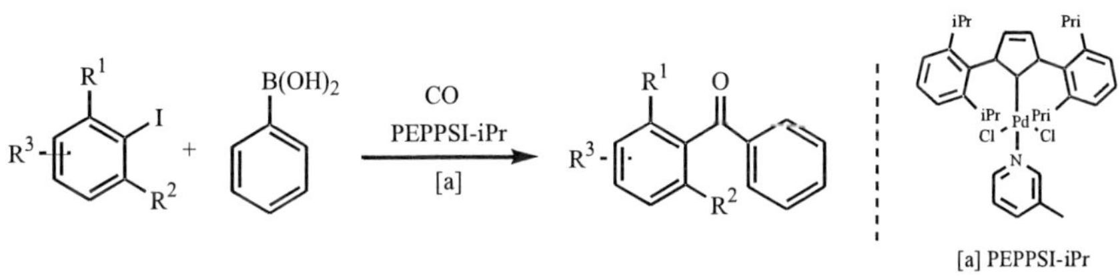

Scheme 14. PEPPSI-iPr catalyst for carbonylative Suzuki reaction.

Carbonylative Stille Coupling

The Stille reaction was discovered in 1977 by John Kenneth Stille. Nearly 50% of all cross-coupling reactions published in 1992 used Stille concept of C-C bond formation. The reaction continues to be exploited industrially, especially for pharmaceuticals. The Carbonylative Kosugi-Migita-Stille reaction, commonly known as the carbonylative Stille coupling, is the palladium-catalysed coupling of carbon monoxide and vinylic systems using organostannanes.

The first heterogeneous carbonylative Stille coupling reaction of organostannanes with aryl iodides under an atmospheric pressure of carbon monoxide has been achieved by Mingzhong Cai et al. in presence of a catalytic amount of an MCM-41-supported bidentate phosphine palladium(0) complex [MCM-41-2P-Pd(0)] [49]. The high catalytic activity and excellent reusability of the MCM-41-2P-Pd(0) complex make them a highly attractive supported palladium catalyst for the synthesis of diverse libraries of compounds (Scheme 15).

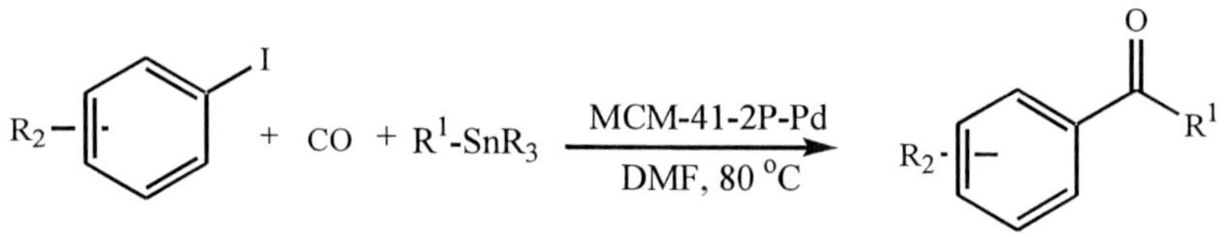

Scheme 15. Carbonylative Stille coupling reaction using MCM-41-2P-Pd catalyst.

CONCLUSION

This review summarized the recent developments in the area of palladium-catalyzed carbonylative coupling reactions of aryl halides and related starting materials in heterogeneous media. Since the original work of Heck and co-workers, various catalytic carbonylation reactions have been developed over the past decades, and nowadays ranges of palladium catalysts are available for these transformations. Thus an impressively diverse range of efficient catalytic systems based on palladium have been developed during past years that prove practical importance of carbonylative coupling reactions.

Heterogeneous catalysis and their application is an expanding field of study that continues to thrive on outstanding laboratory results that have even more significance with the availability of the types of scale-up systems used in processing. Compared with the past, there is now far greater contact and cooperation between the scientific disciplines interested in developing greener and economical protocols for chemical transformations. However, having a plethora of new synthetic catalytic methods for carbonylative coupling reactions, the development of asymmetric catalysts for carbonylation reactions and carbonylative C-H bond activation needs to be explored.

Again most of the carbonylation reactions are reported with palladium catalysts however, these catalysts can be replaced by cheaper and easily available metal catalysts such as Cu, Fe etc. Thus, carbonylation reactions using such catalysts are indeed an area yet to discover.

REFERENCE

[1] Rothenberg, G. (2007). *Catalysis concept and green application.*

[2] Parshall, G. W. (1980). *Homogeneous Catalysis*, John Wiley and sons, New York.

[3] Beller, M., Eckert, M. (2000). Amidocarbonylation: an efficient route to amino acid derivatives. *Angew. Chem., Int. Ed.*, 39, 1010-1027.

[4] Colquhoun, H. M., Thompson, D. J., Twigg, M. V. (1991). *Carbonylation, direct synthesis of carbonyl compounds*, Plenum Press: New York.

[5] Skoda- Foldes, R., Kollar, L. (2002). Synthetic applications of palladium catalysed carbonylation of organic halides. *Curr. Org. Chem.*, 6, 1097-1119.

[6] Kollar, L. (2008). *Modern carbonylation methods*, Wiley-VCH.

[7] Chowdhury, C., Kundu, N. (1999). Studies on copper(I) catalysed cross-coupling reactions : A convenient and facile method for the synthesis of diversely substituted α,β-acetylenic ketones. *Tetrahedron*, 55, 7011-7016.

[8] Logue, M. W., Teng, K. (1982). Palladium-catalyzed reactions of acyl chlorides with (1-alkynyl)tributylstannanes. A convenient synthesis for 1-alkynyl ketones. *J. Org. Chem.* 47, 2549-2553.

[9] Liu, J., Chen, J., Xia, C. (2008). A simple and efficient recyclable phosphine-free catalytic system for alkoxycarbonylation and carbonylative Sonogashira coupling reactions of aryl iodides. *Journal of Catalysis,* 253, 50–56.

[10] Liu, J., Peng, X., Sun, W., Zhao, Y., Xia, C. (2008). Magnetically separable Pd catalyst for carbonylative Sonogashira coupling reactions for the synthesis of α,β-alkynyl ketones. *Organic Letters*, 10, 3933 – 3936.

[11] Genelota, M., Dufaudb, V., Djakovitch L. (2011). Heterogeneous metallo-organocatalysis for the selective one-pot synthesis of *2*-benzylidene-indoxyl and *2*-phenyl-4-quinolone. *Tetrahedron,* 67, 976-981.

[12] Liang, B., Huang, M., You, Z., Xiong, Z., Lu, K., Fathi, R., Chen, J., Yang, Z. (2005). Pd-Catalyzed copper-free carbonylative Sonogashira reaction of aryl iodides with alkynes for the synthesis of alkynyl ketones and Flavones by using water as a solvent. *J. Org. Chem.*, 70, 6097-6100.

[13] Wang, Y., Liu, J., Xia, C. (2011). Cross-linked polymer supported palladium catalyzed carbonylative Sonogashira coupling reaction in water. *Tetrahedron Letters*, 52, 1587–1591.

[14] Rahman, T., Fukuyama, T., Kamata, N., Sato, M., Ryu, I. (2006). Low pressure Pd-catalyzed carbonylation in anionic liquid using a multiphase microflow system. *Chemical Communications*, 21, 2236–2238.

[15] Kantam, M. L., Chowdari, N. S., Rahman, A., Choudary, B. M. (1999). MCM-Silylamine Pd(II) complex: a heterogeneous catalyst for selective azide reductions. *Synlett,* 9, 1413-1414.

[16] Zhou, J. M., Zhou, R. X., Mo, L. Y., Zhao, S. F., Zheng, X. M. (2002). MCM-41 supported aminopropylsiloxane palladium(0) complex: a highly active and stereoselective catalyst for Heck reaction. *J. Mol. Catal. A: Chem.,* 178, 289-292.

[17] Mehnert, P. C., Weaver, D. W., Ying, J. Y. (1998). Heterogeneous Heck catalysis with Palladium-Grafted molecular sieves. *J. Am. Chem. Soc.*, 120, 12289-12296.

[18] Mukhopadhyay, K., Sarkar, B. R., Chaudhari, R. V. (2002). Anchored Pd complex in MCM-41 and MCM-48: novel heterogeneous catalysts for hydrocarboxylation of aryl olefins and alcohols. *J. Am. Chem. Soc.,* 124, 9692-9693.

[19] Kresge, C. T., Leonowicz, M. E., Roth, W. J., Vartuliand, J. C., Beck, J. S., (1992). Ordered mesoporous molecular sieves synthesized by a liquid-crystal template mechanism. *Nature*, 359,710-712.

[20] Hao, W., Sha J., Sheng S., Cai M. (2009). The first heterogeneous carbonylative Sonogashira coupling reaction catalyzed by MCM-41-supported bidentate phosphine palladium(0) complex. *Journal of Molecular Catalysis A: Chemical*, 298, 94–98.

[21] Olah, G. A. (1973). *Friedel–Crafts chemistry*, Wiley, New York.

[22] Yamamoto, Y., Kohara, T., Yamamoto, A. (1976). Selective formation of ketone, diketone and aldehyde by the CO insertion into nickel–alkyl bonds of dialkylnickel complexes. *Chem. Lett.* 5, 1217–1220.

[23] Hatanaka, Y., Fukushima S., Hiyama, T. (1992). Carbonylative coupling reaction of organofluorosilanes with organic halides promoted by fluoride ion and palldium catalyst. *Tetrahedron,* 48, 2113–2119.

[24] Bumagin, N. A., Ponomaryov, A. B., Beletskaya, I. P. (1985). Ketone synthesis via palladium-catalyzed carbonylation of organoaluminium compounds. *Tetrahedron Lett.*, 26, 4819–4822.

[25] Ceccarelli, S., Piarulli, U., Gennari, C. (2000). Effect of ligands and additives on the palladium-promoted carbonylative coupling of vinyl stannanes and electron-poor enol triflates. *J. Org. Chem.*, 65, 6254–6256.

[26] Silbestri, G. F., Bogel-Masson, R., Lockhart, M. T., Chopa, A. B. (2006). A catalyst-free synthesis of asymmetric diaryl ketones from aryltins. *J. Organomet. Chem.*, 691, 1520-1524.

[27] Neumann, W. P., Hillgrtner, H., Baines, K. M., Dicke, R., Vorspohl, K., Kowe, U., Nussbeutel, U. (1989). New ways of genterating organotion reactive intermediates for organic synthesis. *Tetrahedron,* 45, 951-960.

[28] Labadie, J. W., Stille, J. K. (1983). Mechanisms of the palladium-catalyzed couplings of acid chlorides with organotin reagents. *J. Am. Chem. Soc.,* 105, 6129-6137.

[29] Ishiyama, T., Kizaki, H., Miyaura, N., Suzuki, A. (1993). Synthesis of unsymmetrical biaryl ketones via palladium-catalyzed carbonylative cross-coupling reaction of arylboronic acids with iodoarenes. *Tetrahedron Lett.,* 47, 7595-7598.

[30] Khedkar, M. V., Tambade, P. J., Qureshi, Z. S., Bhanage, B. M. (2010). Pd/C: an efficient, heterogeneous and reusable catalyst for phosphine-free carbonylative Suzuki coupling reaction of aryl and heteroaryl iodides. *Eur. J. Org. Chem.*, 6981–6986.

[31] Cai, M., Peng, J., Hao, W., Din, G. (2011). A phosphine-free carbonylative cross-coupling reaction of aryl iodides with aryl boronic acids catalyzed by immobilization of palladium in MCM-41. *Green Chem.,* 13, 190–196.

[32] Herrmann, W. A. (2002). *N*-Heterocyclic carbenes: A new concept in organometallic catalysis. *Angew. Chem. Int. Ed.,* 41, 1290-1309.

[33] Weskamp, T., Kohl, F. J., Hieringer, W., Gleich, D., Herrmann, W. A. (1999). Highly active Ruthenium catalysts for olefin metathesis: the synergy of *N*-heterocyclic carbenes and coordinatively labile ligands. *Angew. Chem. Int. Ed.,* 38, 2416-2419.

[34] Schwarz, J., Böhm, V. P. W., Gardiner, M. G., Grosche, M., Herrmann, W. A., Hieringer, W., Raudaschl-Sieber, G. (2000). Polymer-supported carbene complexes of palladium: well-defined, air-stable, recyclable catalysts for the Heck reaction. *Chem. Eur. J.,* 6, 1773-1780.

[35] Lin, K., Song, M., Cai, M., Hao, X., Wu, Y. (2003). Polymer (fiber)-supported palladium catalyst containing imidazolinyl rings and its application to the Heck reaction. *Tetrahedron Lett.* 44, 3955-3957.

[36] Stevens, P. D., Fan, J., Gardimalla, H. M. R., Yen, M., Gao, Y. (2005). Superparamagnetic nanoparticle supported catalysis of Suzuki cross-coupling reactions. *Org. Lett.,* 7, 2085-2088.

[37] Gambs, C., Dickerson, T. J., Mahajan, S., Pasternack, L. B., Janda, K. D. (2003). High-resolution diffusion-ordered spectroscopy to probe the microenvironment of Jandajel and Merrifield resins. *J. Org. Chem.*, 68, 3673-3678.

[38] Luo, Z., Zhang, Q., Oderaotoshi, Y., Curran, D. P. (2001). Fluorous mixture synthesis: a fluorous-tagging strategy for the synthesis and separation of mixtures of organic compounds. *Science,* 1766-1769.

[39] Dinh, L. V., Gladysz, J. A. (2004). Convenient syntheses of "heavy fluorous" cyclopentadienes and cyclopentadienyl complexes with three to five ponytails. *Chemical Communications*, 998-999.

[40] Harned, A. M., He, H. S., Toy, P. H., Flynn, D. L., Hanson, P. R. (2004). Multipolymer solution-phase reactions: application to the mitsunobu reaction. *J. Am. Chem. Soc.*, 127, 52–53.

[41] Enholm, E. J., Gallagher, M. E. (2001). Free radical reactions on soluble supports from ring-opening metathesis. *Org. Lett.*, 3, 3397–3399.

[42] Bergbreiter, D. E., Osburn, P. L., Smith, T., Li, C., Frels, J. D. (2003). Using soluble polymers in latent Biphasic systems. *J. Am. Chem. Soc.,* 125, 6254-6260.

[43] Hebel, A., Haag, R. (2002). Polyglycerol as a high-loading support for boronic acids with application in solution-phase Suzuki cross-couplings. *J. Org. Chem.*, 67, 9452-9455.

[44] Zheng, S., Xu, L., Xia, C. (2007). Highly efficient *N*-Heterocyclic carbene–palladium complex catalyzed multicomponent carbonylative Suzuki reaction: novel practical synthesis of unsymmetric aryl ketones. *Appl. Organomet. Chem.*, 21,772-776.

[45] Andrus, M. B., Ma, Y., Zang, Y., Song, C. (2002). Palladium–imidazolium-catalyzed carbonylative coupling of aryl diazonium ions and aryl boronic acids. *Tetrahedron Lett.,* 43, 9137-9140.

[46] Maerten, E., Hassouna, F., Bonnaire, S-C., Mortreux, A., Carpentier, J-F., Castanet, Y. (2003). Direct synthesis of benzoylpyridines from chloropyridines via a palladium-carbene catalyzed carbonylative Suzuki cross-coupling reaction. *Synlett,* 12, 1874-1876.

[47] Qureshi Z. S., Deshmukh, K. M., Tambade, P. J., Bhanage, B. M. (2011). A simple, efficient, and recyclable phosphine-free catalytic system for carbonylative Suzuki coupling reaction of aryl and heteroaryl iodides. *Synthesis,* 2, 243-250.

[48] O'Keefe, M., Simmons, N., Martin S. F. (2008). Carbonylative cross-coupling of ortho-disubstituted aryl iodides. Convenient synthesis of sterically hindered aryl. *Org. Lett.,* 10, 5301–5304.

[49] Cai, M., Zheng, G., Ding, G. (2009). The first heterogeneous carbonylative Stille coupling of organostannanes with aryl iodides catalyzed by MCM-41-supported bidentate phosphine palladium(0) complex. *Green Chem.,* 11, 1687–1693.

In: Carbon Monoxide: Sources, Uses and Hazards ISBN:978-1-61942-055-7
Editors: D. DiLoreto et al.

Chapter 5

EVALUATION OF CARBOXYHEMOGLOBIN (CO-HB) BY OXIMETER IN FORENSIC PRACTICE

Hiroshi Kinoshita*, ***Naoko Tanaka, Mostofa Jamal, Mitsuru Kumihashi and Kiyoshi Ameno***
Department of Forensic Medicine, Faculty of Medicine,
Kagawa University, Miki, Japan

ABSTRACT

Carbon monoxide (CO) is the leading cause of poisoning death in Japan. As CO is generated by the imperfect combustion of various organic compounds, fire and other causes, measurement of the saturation of carboxyhemoglobin (CO-Hb) in blood is important in forensic practice. Quantitation of CO-Hb is usually performed by various methods such as gas chromatography or spectrophotometry. In the present study, we determined CO-Hb saturation in blood samples obtained from fire-related deaths and CO poisoning cases using a portable oximeter AVOX 4000 (AVOX). The CO-Hb was also measured by the conventional spectrophotometric method. There was a good correlation between the

* All correspondence concerning this paper should be addressed to: Dr. H. Kinoshita, Department of Forensic Medicine, Faculty of Medicine, Kagawa University, 1750-1, Ikenobe, Miki, Kita, Kagawa, 761-0793, Japan. TEL: +81-87-891-2140. FAX: +81-87-891-2141e-mail: kinochin@med.kagawa-u.ac.jp

values obtained by AVOX and by the conventional method, within the range of 10-75% of CO-Hb. In over 75% of CO-Hb by the conventional method, the values were indicated "75%<" by AVOX. AVOX is suitable for forensic practice as a new tool by easy handling with a small sample in comparison with conventional methods.

Keywords: Carbon monoxide, Poisoning, Oximeter, Carboxyhemoglobin

INTRODUCTION

Carbon monoxide (CO) is an odorless and colorless gas, which is generated by the imperfect combustion of various organic compounds, fire and other causes [1]. CO is the leading cause of poisoning death in Japan [2, 3]. The annual number of victims by CO poisoning are about 4000 in Japan, and most of them are fire-related death or suicide. CO poisoning is of relatively high incidence in cases of forensic practice [4].

Measurement of carboxyhemoglobin (CO-Hb) concentration in blood is important for the diagnosis in cases of CO poisoning or death by fire. Quantitation of CO-Hb is usually performed by gas chromatography or spectrophotometry [5-8]. Oximeter is routinely used in clinical practice [9,10] and it also applies in forensic practice [11-21].

In this paper, we describe that the usefulness of oximeter AVOX 4000 (AVOX, International Technidyne Corporation, NJ, USA), which is developed for the near-patient testing in the intensive care unit [22], for CO-Hb measurement in daily forensic practice.

MATERIALS AND METHODS

Femoral venous blood, left and right heart blood were routinely examined at medicolegal autopsy from 20 victims (60 samples) exposed to fire or CO gas. The subjects contained 10 males and 10 females, 24-89 years of age. Their post-mortem intervals ranged from approximately 4 to 72 hrs.

Measurement of CO-Hb in blood was performed by both AVOX and the conventional spectrophotometric method. The operation of AVOX and the preparation of the samples were in accordance with the manufacturer's specifications [21,23]. In brief, approximately 50 μl of sample blood were filled into the cuvette, and inserted into the AVOX. The sample is analyzed automatically within 10 seconds. Conventional spectrophotometric measurement of CO-Hb was performed in accordance with the previous report [7].

RESULTS AND DISCUSSION

The CO-Hb concentration in blood using AVOX were 15.3 to 75%< in left heart blood, 14.3 to 75%< in right heart blood and 12.1 to 75%< in femoral venous blood, respectively. As the operating range of AVOX is 0-75% for CO-Hb, we could not obtain the accurate value of CO-Hb by AVOX when the concentration of CO-Hb is over 75%. We have subdivided those samples according to the CO-Hb concentration by AVOX as follows; (a) samples with CO-Hb higher than 75% (n=29), and (b) samples with 10-75% of CO-Hb (n=31), and compared with the data of the conventional method.

In all of the samples of group (a), the values of the CO-Hb concentration by the conventional method were higher than 75% (Figure 1).

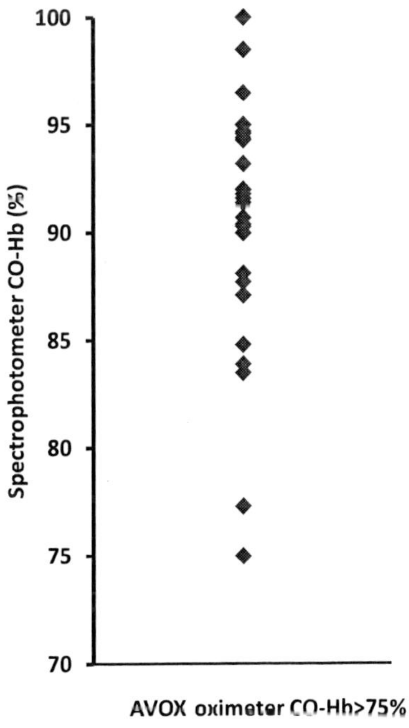

Figure 1. Carboxyhemoglobin (CO-Hb) values of sample (a) by the conventional spectrophotometric method.

The relationship of CO-Hb value between the AVOX and conventional method in group (b) are shown in Figure 2. The values by AVOX were almost equivalent to those by the conventional method. (regression equation of Y=0.9235X+5.2914; Y: CO-Hb by the conventional method, X: CO-Hb by AVOX, correlation coefficient, r=0.9627). It demonstrated an excellent correlation in a range of 10-75% of CO-Hb.

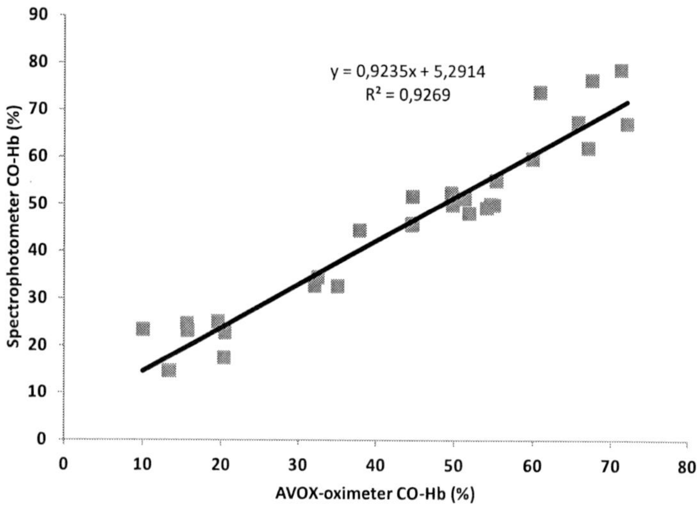

Figure 2. Relationship of carboxyhemoglobin (CO-Hb) concentrations (%) between AVOX and the conventional spectrophotometric method.

We investigated the arterial-venous difference of CO-Hb concentration. We used 9 cases of blood samples that all CO-Hb concentrations were less than 75% by AVOX. The relationship of CO-Hb concentrations between the left and right heart blood is shown in Figure 3. The CO-Hb levels in left heart blood were almost correlated with those in right heart blood. (regression equation of Y=0.9826X-0.2724; Y: CO-Hb in the right heart blood, X: CO-Hb in the left heart blood, correlation coefficient, r=0.9985). The relationship of CO-Hb concentrations between the left heart blood and femoral venous blood is shown in Figure 4. The CO-Hb levels in left heart blood were similar to those in femoral venous blood. (regression equation of Y=1.0728X-5.5227; Y: CO-Hb in the femoral venous blood, X: CO-Hb in the left heart blood, correlation coefficient, r=0.9883). No remarkable differences of CO-Hb concentration were observed between arterial and venous blood in a clinical and experimental study [10, 24].

This may be due to that the blood circulation may be continued following the cessation of respiration in CO poisoning or fire-related case [25]. This data suggests that site difference of CO-Hb concentration is negligible at the range of less than 75% of CO-Hb.

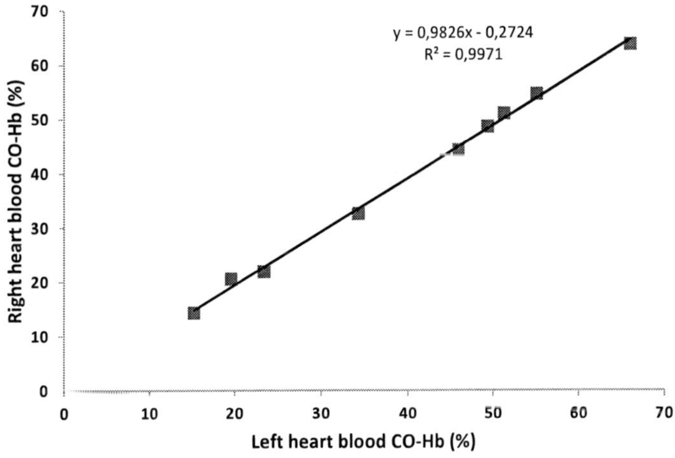

Figure 3. Relationship of carboxyhemoglobin (CO-Hb) concentrations (%) between the left and right heart blood.

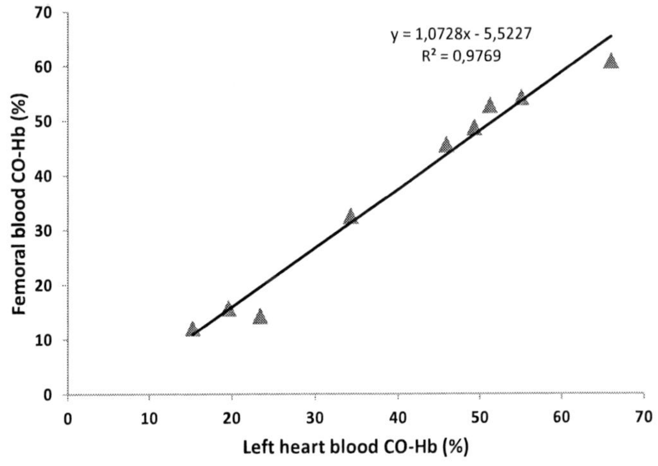

Figure 4. Relationship of carboxyhemoglobin (CO-Hb) concentrations (%) between the left heart blood and femoral venous blood.

The AVOX is a portable oximeter, which uses seven wavelengths for the measurement of various hemoglobin species [22,26,27]. It automatically analyzes the contents of total hemoglobin, oxyhemoglobin, CO-Hb, methemoglobin and oxygen contents in a small amount of samples (50μl) within a few seconds. AVOX is suitable for the forensic practice as a new method since there are many advantages such as its portability, no necessity of sample preparation, easy handling and only needs a small sample which reduces the risk of infection in comparison with conventional methods.

REFERENCES

[1] Baselt RC. (2000) Carbon monoxide. In: Baselt RC (Ed), Disposition of toxic drugs and chemicals in man, (5th ed). pp133-136. Foster City, CA: Chemical Toxicology Institute.

[2] Ministry of Health, Labor and Welfare. (2007) Vital statistics of Japan.

[3] National Research Institute of Police Science. (2011) Annual case reports of drugs and toxic poisoning in Japan (No53), Kashiwa: National Police Agency.

[4] Investigation committiee of the Medico-Legal Society of Japan. (1991) Report on medico-legal data from the massive-investigation performed by the Medico-Legal Society of Japan – A statistical study of death by poisoning. *Nihon Houigaku Zasshi*, 45, 258-262.

[5] Akane A, Fukui Y. (1985) A review on the development of carboxyhemoglobin analysis. *Res. Pract. Forens Med.*, 28, 185-190.

[6] Sato K. (2005) Carbon monoxide. In: Suzuki O and Watanabe K (Eds), Drugs and poisons in humans. A handbook of practical analysis. pp 91-99. Berlin, Heidelberg, New York: Springer-Verlag.

[7] The Pharmaceutical Society of Japan (1992) Carbon Monoxide. In: Standard methods of chemical analysis in poisoning –with commentary, (4th ed). pp57-68. Tokyo, Nanzando.

[8] Working group on Forensic Toxicology (2006) Carbon Monoxide. In: Manual for forensic toxicology analysis of the Japanese Society of Legal Medicine, (2nd ed). pp22-25. Tokyo, *Japanese Society of Legal Medicine.*

[9] Mahoney JJ, Vreman HJ, Stevenson DK, Van Kessel AL. (1993) Measurement of carboxyhemoglobin and total hemoglobin by five specialized spectrophotometers (CO-oximeters) in comparison with reference methods. *Clin. Chem.*, 39, 1693-1700.

[10] Touger M, Gallagher EJ, Tyrell J. (1995) Relationship between venous and arterial carboxyhemoglobin levels in patients with suspected carbon monoxide poisoning. *Ann. Emerg. Med*, 25, 481-483.

[11] Okada M, Okada T, Ide K. (1985) Utilization of a CO-oximeter in the medico-legal field. *Nihon Houigaku Zasshi*, 39, 318-325.

[12] Ohshima T, Takayasu T, Nishigami J, LinZ, Kondo T, Nagano T. (1992) Application of hemoglobin analysis by CO-oximeter to medico-legal practice with special reference to diagnosis of asphyxia. *Nihon Houigaku Zasshi*, 46, 382–388.

[13] Higuchi T, Noguchi K, Maeda H. (1992) An evaluation of analyzed data of hemoglobin derivatives by CO-oximeter in medico-legal autopsy. *Nihon Houigaku Zasshi*, 46, 416–418.

[14] Maeda H, Fukita K, Oritani S, Nagai K, Zhu BL. (1996) An evaluation of post-mortem oxymetry in fire victims. *Forensic. Sci. Int.*, 81, 201–209.

[15] Oritani S, Nagai K, Zhu BL, Maeda H. (1996) Estimation of carboxyhemoglobin concentrations in thermo-coagulated blood on a CO-oximeter system: an experimental study. *Forensic. Sci. Int.*, 83, 211–218.

[16] Maeda H, Fukita K, Oritani S, Ishida K, Zhu BL. (1997) Evaluation of post-mortem oxymetry with reference to the causes of death. *Forensic Sci Int*, 87, 201–210.

[17] Shimizu K, Mizukami H, Fukushima T, Sasaki M, Shiono H. (1998) Use of a CO-oximeter for forensic diagnosis of hypothermia. *Nihon Houigaku Zasshi,* 52, 196–201.

[18] Brehmer C, Iten PX. (2003) Rapid determination of carboxyhemoglobin in blood by Oximeter. *Forensic. Sci. Int*, 133, 179–181.

[19] Watanabe N. (2003) Medico-legal application of hemoglobin analysis using CO-oximeter. *Hokkaido Igaku Zasshi*, 78: 557-566.

[20] Olson KN, Hillyer MA, Kloss JS, Geiselhart RJ, Apple FS. (2010) Accident or arson: Is Co-oximetry reliable for carboxyhemoglobin measurement post-mortem? *Clin. Chem,* 56: 515-520.

[21] Tanaka N, Ameno K, Jamal M, Ohkubo E, Kumihashi M, Kinoshita H. (2010) Application of oximeter AVOX 4000 for the determination of CO-Hb in forensic practice. *Res. Pract. Forens. Med,* 53, 39-43.

[22] Gong AK. (1995) Near-patient measurement of methemoglobin, oxygen saturation, and total hemoglobin: Evaluation of a new instrument for adult and neonatal intensive care. *Crit. Care Med,* 23, 193-201.

[23] AVOXimeter 4000 whole blood oximeter operator's manual. NJ, USA : International Technidyne Corp.

[24] Lopez DM, Weingarten-Arams JS, Singer LP, Conway EE. (2000) Relationship between arterial, mixed venous, and internal jugular carboxyhemoglobin convcentrations at low, medium, and high concentrations in a piglet model of carbon monoxide toxicity. *Crit. Care. Med.,* 28, 1998-2001.

[25] Miyazaki T, Kojima T, Yashiki M, Chikasue F, Iwasaki Y. (1992) Interpretation of COHb concentrations in the left and right heart blood of cadavers. *Int. J. Leg. Med.,* 105, 65-68.

[26] Brown LJ. (1980) A new instrument for the simultaneous measurement of total hemoglobin, % oxyhemoglobin, % carboxyhemoglobin, % methemoglobin, and oxygen content in whole blood. *IEEE Trans. Biomed. Eng*, 27, 132-139.

[27] Bailey SR, Russell EL, Martinez A. (1997) Evaluation of the AVOXimeter: precision, long-term stability, linearity, and use without heparin. *J. Clin. Monit,*13,191-198.

In: Carbon Monoxide: Sources, Uses and Hazards ISBN:978-1-61942-055-7
Editors: D. DiLoreto et al.

Chapter 6

RETROSPECTIVE ASSESSMENT OF CARBON MONOXIDE INTOXICATION RISK FACTORS IN CIUDAD JUÁREZ, MEXICO

Teresa Montoya[1], Patrick L. Gurian[2], Verónica Corella-Barud[1] and Zuber D. Mulla[3]

[1] Center for Environmental Resource Management, University of Texas at El Paso, El Paso, TX, USA

[2] Department of Civil, Architectural, and Environmental Engineering, Drexel University, Philadelphia, PA, USA

[3] Texas Tech University Health Sciences Center, Paul L. Foster School of Medicine, Obstetrics and Gynecology, El Paso, TX, USA

ABSTRACT

Between 1992 and 2004, 1,381 people living in Ciudad Juárez, México, were intoxicated, and 167 more have lost their lives to carbon-monoxide poisoning. This study examines 91 incidents of carbon-monoxide intoxication from October to February 2003-2004. Gas heaters were identified as the cause of 59% of the incidents. Carbon-monoxide intoxication incidents are associated with colder temperatures on both a monthly average and a daily basis. A neighborhood-level analysis suggests that households in the middle of the income spectrum are disproportionately at risk. An exploratory comparison of risk factors between the case data and baseline data from three neighborhoods found that gas heaters are associated with higher risks than wood or charcoal heaters. In 32% of the intoxication incidents, doors and windows of the home were sealed with plastic, newspaper or other materials. While differences are noted in the relative risks for heaters produced by different manufacturers, these differences are likely not applicable to current models as the manufacturers have introduced new designs. Interventions supported by this research include efforts to encourage proper ventilation of homes, the replacement of older heaters with newer models that have oxygen depletion sensors, the use of carbon monoxide detectors, and turning off heaters while sleeping.

Background

Between 1992 and 2004, 1,381 people living in Ciudad Juárez, México, were intoxicated and 167 more lost their lives to carbon-monoxide (CO) poisoning (*Diario de Juárez* 2004). In the months from October to February 2003-2004, 212 people were intoxicated by CO and 11 more died. The danger is greatest during the winter season when many families in Ciudad Juárez use un-vented combustion heaters.

CO exposures are associated with many adverse health effects ranging from headaches, dizziness, nausea, loss of consciousness, coma and even death. Chronic exposure lasting more than a week can result in a recognizable syndrome of persistent neurological sequelae (Bayer, et. al. 1998). It is common that individuals suffering from CO poisoning are unaware of their exposure because symptoms are similar to those associated with viral illness or clinical depression (Raub, et. al. 2000). Worldwide, the exact number of carbon monoxide cases is not known since it is grossly under-reported or misdiagnosed by medical professionals. Nevertheless, it is believed to account for more than one-half of the fatal poisonings reported in many countries (Raub, et. al. 2000). In the United States, it is the most common cause of fatal poisoning, accounting for approximately 1100 accidental deaths annually (Ralston and Hampson 2000).

Previous Research

Previous studies in the U.S. have already identified some common risk factors for unintentional CO poisoning. While CO poisoning is a year-round problem, the incidence of cases increases during the winter months because of the use of heaters or other fuel-burning appliances along with lack of ventilation (Randolph, et. al. 2000). Liu et al. (2000) observed a sharp increase in the number of deaths per month in California during the winter, with December as the month with the greatest number of deaths. Cobb and Etzel (1991), in their study of case reports from across the U.S. found that the yearly frequency of deaths in January was four times the frequency in July. In the same study, the highest death rates occurred in states with cold climates, such as Alaska.

Gas heating appliances were associated with a 45-fold higher risk of fatal unintentional CO poisoning in California relative to central, forced-air gas heating (Liu, et. al. 2000). Charcoal-burning appliances appeared to be associated with elevated rates of CO intoxication, although sufficient data were not available to calculate an odds ratio for this risk factor. In Washington State, charcoal briquettes were the most common source of CO poisoning (Ralston, and Hampson 2000).

Less information is available for CO poisoning in northern Mexico where Ciudad Juárez is located. A previous study (Graham et al. 2004) of health-effects and environmental risks associated with growth of peri-urban areas in Ciudad Juárez found that elevated levels of carbon monoxide were common inside households. Instantaneous measurements for carbon monoxide were made in 299 low-income households and over two-thirds were found to exceed the 9 ppm 8-hour exposure level established by the U.S. Environmental Protection Agency and the World Health Organization. A significant minority exceeded 60 ppm CO. These data, along with the numerous case reports of intoxications show that CO intoxication represents a serious problem in Ciudad Juárez, and yet no previous studies have identified the specific risk factors for CO poisoning in this region. Risk factors in Ciudad Juárez are likely to differ significantly from risks reported in previous U.S. studies given differences in culture, income, climate, and housing infrastructure. This study is an initial examination of the case-reports in Ciudad Juarez in order to understand regional patterns in CO intoxication. The spatial and temporal distribution of intoxication incidents is described, common factors in case reports identified, and the frequency of risk factors among the cases are compared with frequencies of these factors among non-cases in databases from previous studies. The case reports are from neighborhoods throughout Ciudad Juárez while the baseline data are from a subset of three neighborhoods of Ciudad Juárez. This is a source of potential bias in the study, as differences between the prevalence of risk factors may reflect differences between the three neighborhoods and the rest of Ciudad Juárez that are unrelated to CO poisoning. For this reason, the terminology of "case-control study" is avoided throughout the paper. However, the investigators believe that the cases and baseline data come from essentially similar populations, namely lower-income residents of Ciudad Juárez. The results are presented here as they represent the most complete data yet available on an important health risk and may serve to identify potential risk factors for more rigorous follow up studies.

METHODS

A series of 91 reports of CO poisoning incidents in the winter (October-February) 2003-2004 was obtained from the Ciudad Juárez office of Protección Civil. An incident was defined as a household where its occupants presented one or more of the symptoms of acute CO intoxication including headaches, nausea, fainting, etc. The analysis was conducted on the incident level rather than the individual case level, because a single cause was associated with each incident (i.e., individual case data are not independent). These incidents included both fatalities and non-fatalities from households that contacted the Ciudad Juárez emergency number (060) and reported a suspected CO poisoning. Incidents were later verified by the responding firefighters and documented in written reports. Poisonings were reported from 64 different neighborhoods. Of the 91 incidents, 85 reported a reason for the intoxication, while in 6 incidents the reason was unknown.

Baseline frequency data were obtained from two previous studies in Ciudad Juárez. The information referred to throughout this article as baseline Database 1 came from a survey of peri-urban households in Ciudad Juárez for environmental health risks. Surveys were administered in the winter of 2001-2002 during home visits by researchers (Graham, et. al. 2004). Three neighborhoods were included in the survey; one located in the eastern part of the city, one located in the west, and one located in the south central part of the city. The total number of households in this database was 299, all of low socio-economic status. The information in the other database (Database 2) was obtained from a study performed during the 2003-2004 winter season. The purpose of the study was to monitor indoor levels of CO for 63 households in Ciudad Juárez. The households for this study were also selected from the same peri-urban, low-income communities previously mentioned. Database 2 is smaller than Database 1, but contains more detailed information related to home heating, such as heater manufacturer.

The data obtained from the case reports and baseline databases were entered into Microsoft Excel and analyzed using SPSS statistical software. Crude odds-ratios were calculated and chi-square tests were conducted to evaluate each risk factor.

RESULTS

Demographic Patterns of Cases

The case reports came from 64 different neighborhoods in Ciudad Juárez. The city was divided into an Eastside and Westside taking as a reference the railroad tracks. Low-income settlements lacking basic infrastructure, such as water supply, paved roads, and adequate housing, are present on both sides of the city but are more prevalent on the Westside than the Eastside (Zona Poniente Plan Parcial de Desarrollo 2001). The Westside has 42% of the total population (INEGI 2000) and was identified as the location of 22% of the CO cases. In contrast, the Eastside has 58% of the population and was identified as the location of 63% of the cases. The neighborhood for 16% of the cases was not reported. Even if all the cases from unknown locations were from the Westside, the Eastside still has a disproportionate share of the cases. This suggests that the generally poor housing infrastructure on the Westside may actually be protective against CO poisoning. Many residents of the Westside live in improvised dwellings, often made of pallets or cardboard (Graham et al. 2004). These dwellings would be expected to have high air exchange rates which would limit the extent to which CO could build up indoors.

Socio-economic status was also investigated as a means to help identify the at-risk populations. Unfortunately information on the socioeconomic status of individual households was not included in the case reports. However, the distribution of household income by neighborhood is available (INEGI 2000). The Mexican minimum wage of $45.24 *pesos* per day (980 pesos/month) was taken as a basis of comparison, and the distribution of income, in neighborhoods with reported cases was compared to the income distribution in neighborhoods without reported cases (Figure 1a and Figure 1b). The two distributions are similar, but very poor and very wealthy households appear to be somewhat less common in neighborhoods with reported cases. Only 2% of households in neighborhoods with CO incidents earn less than the full-time minimum wage, compared to 3% of households in other neighborhoods. At the other end of the income scale, 14% of households in neighborhoods with CO intoxications earn more than five times the minimum wage, while 19% do so in neighborhoods without reported CO incidents.

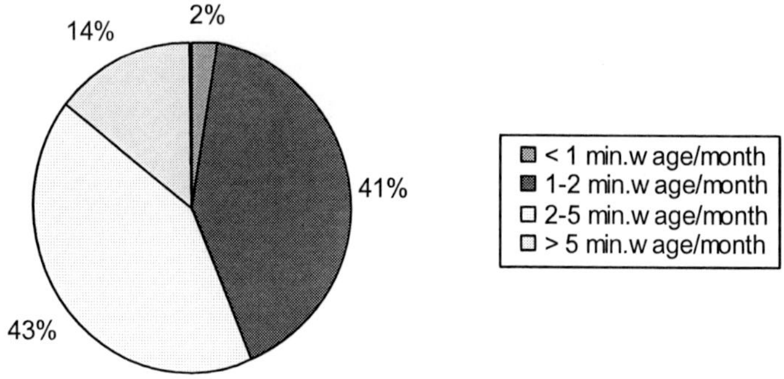

Figure 1a. Income distribution for neighborhoods having one or more reported cases of CO intoxication.

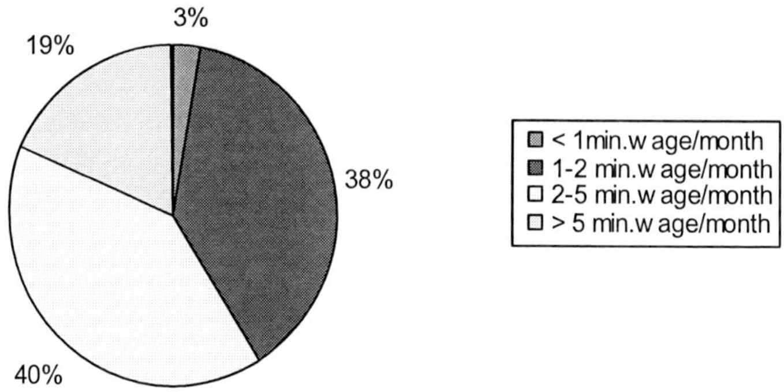

Figure 1b. Income distribution for neighborhoods having one or more households without reported cases of CO intoxication.

Seasonal Trends

The US-México border region of Ciudad Juárez is known for its extreme temperatures. During the winter months, nighttime temperatures can drop well below freezing; during daytime, temperatures frequently reach the 70s°F (National Weather Service 2004). The coldest months in the region are December and January. It is during these two months when the greatest numbers of CO poisoning incidents are seen in Ciudad Juárez.

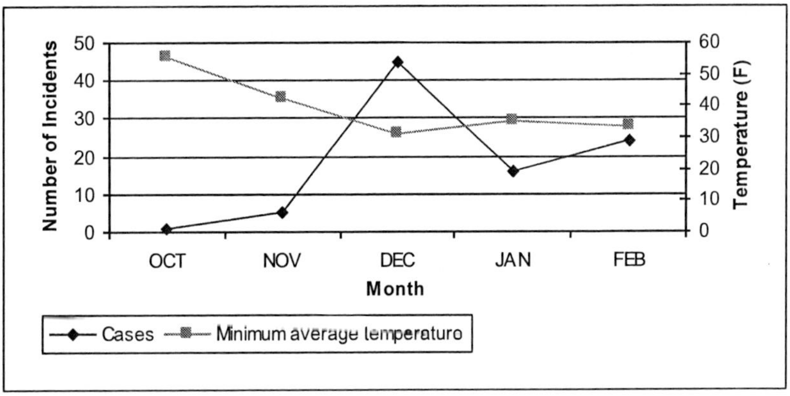

Figure 2. Reported CO cases and minimum average monthly temperature.

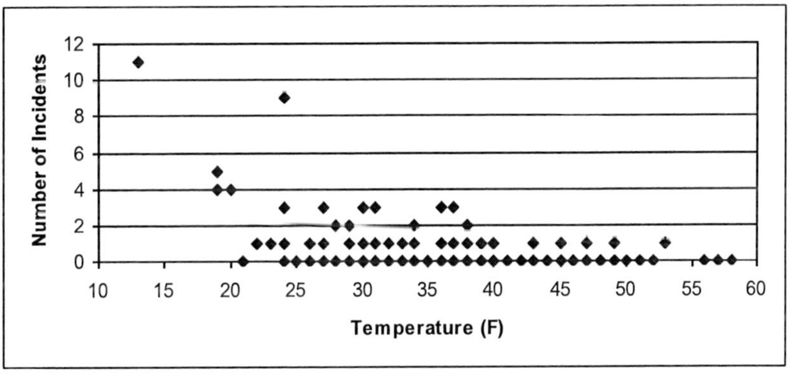

Figure 3. Total number of CO incidents and minimum temperature for each day from October 31, 2003 thru February 29, 2004.

Figure 2 shows that for the 2003-2004 winter season, the month of December recorded the lowest average temperature of 31°F and had the highest number of cases, 45. On the other hand, October, with an average minimum temperature of 56°F, had only one incident reported. This corresponds to the seasonal trend observed in U.S. studies (Liu et al. 2000, Cobb and Etzel 1991). Figure 3 shows the total number of CO poisoning incidents for each day during the time-period from October 31, 2003, when the first CO poisoning of the winter season was registered, through February 29, 2004, as compared to the daily minimum temperature. The graph indicates that the greatest number of cases occurred on the days when the lowest temperatures of the region were registered.

The three coldest nights, one with a minimum temperature of 13°F and two with minimum temperatures of 19°F, accounted for 22% of all the incidents. On nights with a minimum temperature above 40°F, no more than a single incident of CO poisoning took place. The relationship between minimum daily temperature and number of cases (Spearman correlation coefficient = –0.44) is slightly stronger than the relationship between average daily temperature (Spearman correlation = –0.42) and maximum daily temperature (Spearman correlation = –0.35), which suggests that nighttime temperatures may place the greatest role in influencing risk. All three Spearman correlations are significantly different from zero ($p<0.001$).

Common Risk Factors

Reports filed by the Fire Department usually identified a suspected cause of the CO intoxication. As shown in Figure 4 gas heaters accounted for 59% of the incidents. The second largest category (18% of incidents) included all the solid-fuel appliances, such as *braseros* (improvised wood-burning heaters), drums, or grills.

Table 1 summarizes the odds ratios and chi-square calculations comparing the frequency of various risk factors in the case reports and the baseline data. The odds ratio for homes with gas heaters vs. homes without gas heaters was 3.3 with a p-value <0.001, indicating that gas heaters are significantly associated with increased risk. Homes lacking gas heaters included households with central heating, those that used electric, wood or charcoal heaters, those using gas stoves for heat, and those without any heat source. An overall comparison of the safety of gas and electric heaters is not possible on the basis of this study. While electrical heaters would not pose the same CO intoxication risk as gas heaters, they present other risks, such as burns and electrical fires. When comparing gas heaters with solid-fuel heaters (wood and charcoal) the odds ratio was 2.4 with a p-value of 0.009, indicating that gas heaters present significantly greater risks of CO intoxication. This may be because gas heaters are usually un-vented, while some of the wood heaters are vented. Information on how frequently the solid fuel heaters were vented is not available in Database 1 or in the case reports, making it impossible to calculate an odds ratio for this factor. However, 13 of the 15 solid fuel heaters in Database 2 were vented. A second possible explanation of the excess risk associated with gas heaters is that wood heaters will tend to burn down if left overnight, while the gas heaters can burn throughout the night.

Of the 91 cases reported, 54 involved a gas heater and 53 of the gas heaters were reported to have *radiantes* (ceramic bricks that are heated by the flame and subsequently radiate heat to the room). A previous study done to compare air pollutant emissions from portable convective heaters vs. radiant kerosene space heaters found that CO emissions were higher for the space heaters with *radiantes* due to less efficient combustion (Leaderer 1982). Gas heaters are available that do not use *radiantes*. The baseline database did not have complete information on the number of *radiantes* per gas heater, making an odds ratio or chi-square calculation infeasible, but the presence of *radiantes* in almost all of the heaters suggests that this may be a factor deserving attention in future studies.

Figure 5 shows the brand of gas heaters involved in the 54 incidents. Of these, 26 were manufactured by *El Sol* and 10 by *Solmatic*. These companies are located in Ciudad Cuauhtémoc, México and produce heaters of similar design (Apodaca 2005) for the lower end of the market. The two companies recently merged (Santana 2004). 12 gas heaters were from other manufacturers that included 2 from Europe, 3 from the United States, 1 from Chile, and an additional 6 from México. The manufacturer was not reported for 6 incidents. Having an *El Sol* heater, compared to all other brands, is associated with an odds ratio of 3.9 ($p<0.001$).

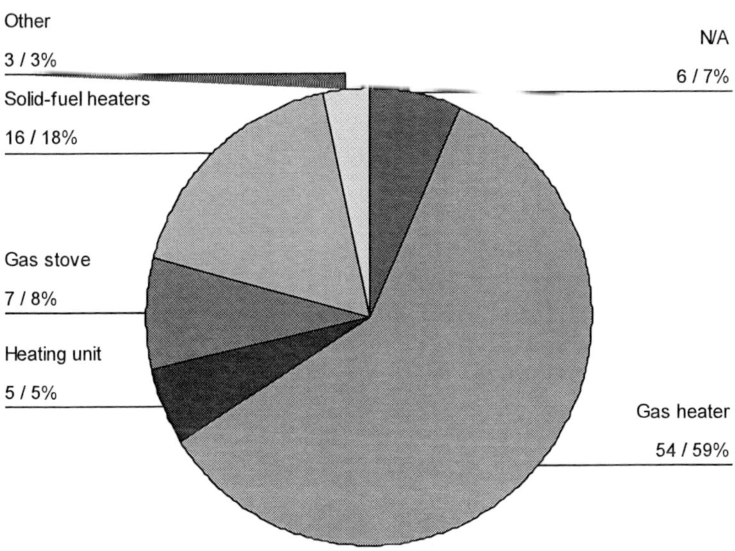

Figure 4. Appliance being used at the time of the incident.

Table 1. Comparison of cases and baseline data

Risk Factor	Referent Group	Source of Baseline Data	Number of Applicable Cases	Prevalence of risk factor among cases	Number of Applicable Baseline Households	Baseline prevalence of risk factor	Odds Ratio	P-Value
Gas heaters	No gas heater present	Database 1	91	59%	299	30%	3.3	<0.001
El Sol heaters	Other gas heaters	Database 2	54	48%	63	19%	3.9	<0.001
Solmatic heaters	Other gas heaters	Database 2	54	19%	63	13%	1.6	0.38
Solmatic heaters	El Sol heaters	Database 2	36	28%	20	40%	0.58	0.35
Gas vs. solid-fuel heaters	Solid-fuel heaters	Database 1	70	77%	154	59%	2.3	0.01
Gas stove for heating	No gas stove for heating	Database 1	91	1%	299	8%	0.12	0.02
Being from out of town	Being born in Juárez	Database 1	41	41%	58	55%	0.58	0.18
Heater on at night	Heater off at night	Database 2	91	16%	63	5%	3.9	0.026
Windows/ doors sealed	Not sealing windows /doors	Database 2	91	32%	63	13%	3.2	0.006

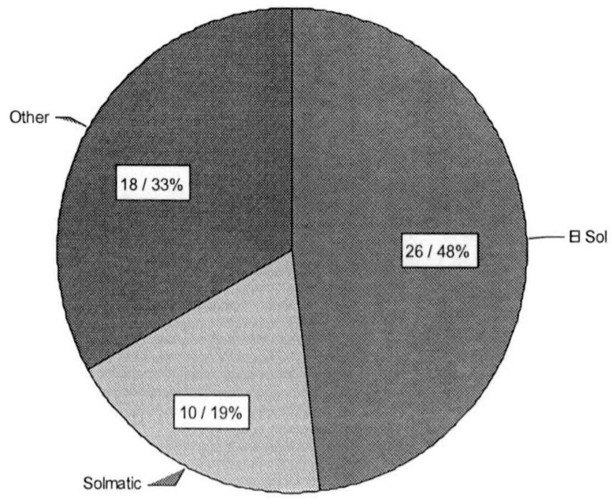

Figure 5. Most common brands of gas heaters among cases.

The odds ratio for *Solmatic* heaters, compared to all other brands, was 1.6, which is suggestive of an elevated risk but not significantly greater than 1. When the risk for *El Sol* heaters is compared to *Solmatic*, the odds ratio is very close to 1 (0.88) and non-significiant (p=0.35), suggesting that there is little difference in risk between the two most common manufacturers.

In 32% of the incidents the houses were reported to be *herméticamente cerradas* (completely sealed) meaning that the doors and windows were covered with plastic, newspaper, or other material as an insulating measure. The odds ratio for these households indicated a significantly elevated risk (odds-ratio of 3.2, p=0.006) when compared to the available baseline data. This practice appears to be unsafe and may be an attractive target for interventional efforts (see Discussion below).

One potential intoxication scenario involves a family that leaves a heater on at night. While at least some families understand that this is an unsafe practice (Focus Group 2004), it may occur either by accident (the family forgets to turn off the heater) or intentionally, perhaps against their better judgment, during a period of severe cold. The excess risk associated with low minimum temperatures appears to support this scenario, but there is no direct information in the case reports as to whether the heater was left on overnight. However, the time of the incident was recorded. For this study all incidents reported as occurring between 10 pm and 9 am were classified as being associated with "nighttime use" (likely overnight use) of a heater. Similarly there was no direct information on whether baseline households left their heaters on overnight, but total hours that the heaters was used at night was recorded. Those households that reported more than eight hours of heater use per night were classified as being in this "nighttime use" risk category. "Nighttime use" as defined in this manner was associated with a significantly elevated risk of intoxication (odds ratio of 3.9 and p=0.026). This practice may be another potentially important target for public education and behavioral interventions.

Another risk factor considered was the use of a gas stove for heating. Among the cases, only 1 reported using a gas stove specifically for heating. In baseline Database 1, 25 households reported using a gas stove for heating. The odds ratio was 0.12 with p=0.015, indicating that gas stoves are associated with a lower risk of intoxication than other heat sources. Liu et al. (2000) reported that gas stoves were unsafe as they accounted for half as many fatalities in California as portable heaters. Liu et al. did not have baseline data on the use of gas stoves for heating making it impossible for them to calculate an odds ratio. In this study, the ratio of incidents involving gas stoves to incidents involving portable heaters is clearly lower (1 to 54 rather than 15 to 33). It may be that portable heaters used in California are safer or that stoves are used differently for heating in the two regions, with Californians being more likely to have a piped gas supply and perhaps using the stoves more intensively for heat.

Previous studies (Graham et al. 2004) have speculated that people from out of town are at greater risk of being intoxicated with CO. In contrast, the local newspaper reports that 60% of intoxicated people were natives of Juárez (*Diario de Juárez* 2004). For this study, households were classified as "out of town" households if all members were born outside of Juarez and natives of Juárez if all members of the household were born in Juárez. Based on these definitions, 26% of the households were natives of Juarez and 17% were from out of town. The rest of the households had families with mixed birthplaces and these were omitted from the odds-ratio calculation. The odds ratio for being from out of town was 0.58 with p=0.18, considered non-significant. Recent migrants to Ciudad Juárez may be less likely to be able to afford gas heaters and may either go without heat or use the less expensive, improvised solid-fuel heaters which appear to be safer than gas heaters (as discussed earlier in this section).

DISCUSSION

This study identified several factors that appear to be associated with excess risk of CO intoxication. The majority of intoxications were associated with small heating appliances that burn either gas or wood. CO intoxication incidents are associated with colder temperatures on both a monthly average and a daily basis. This confirms that intoxications are largely related to winter heating, as reported by previous researchers in other regions (Liu et al. 2000, Randolph, et. al. 2000, Cobb and Etzel 1991). Intoxications are reported throughout the city but appear to somewhat disproportionately affect the Eastside (which has better infrastructure) and households in the middle of the income distribution. A single incident is associated with a malfunctioning water heater and likely comes from a higher-income household, but in general high-income houses that rely on central heating appear to be at decreased risk. Very low-income households may also be at decreased risk, either because they can not afford any form of heating or because they live in poorly insulated houses that have high air exchange rates.

An exploratory comparison of risk factors between the case data and baseline data from three neighborhoods of the city found a number of factors that differed in prevalence between the two groups. These factors may be associated with excess risk of CO intoxication, or may just be indications of differences between the three neighborhoods and the city as a whole. These factors include having a gas heater, and sealing doors and windows for insulation, the use of certain older heaters produced in México, and the use of the appliances at night/overnight. Wood and charcoal heaters appeared to be protective against CO intoxication, either because some of them are vented or because they burn down if left unattended overnight.

As discussed in the Introduction, differences between the incident database and the baseline databases may be due to differences between the three neighborhoods from which the baseline data was taken and the wider population of Ciudad Juárez. For example, the higher rate of use of gas heaters than wood heaters among the cases could simply be an indication of higher socioeconomic status allowing them to opt for the more convenient and more expensive gas heaters. Alternatively, the cases could be from lower income households and therefore more likely to use the lower cost, Mexican-made heaters. Because of these concerns, this study can not conclusively include or exclude possible risk factors. However, the baseline data is intended to be generally relevant to lower-income neighborhoods of Ciudad Juárez, as it was collected from three different neighborhoods with housing types ranging from small cement-block houses to shanties made of pallets or cardboard (Graham et al. 2004).

Given the lack of any more rigorously-controlled studies, the information from this study, if interpreted cautiously, may help to identify the appropriate target and timing of public information campaigns. The results generally support Protección Civil's existing practice of providing information emphasizing safe heating practices during the winter and reinforcing this message with radio and television warnings on particularly cold nights which have a disproportion share of the overall incidents (Figure 3). The results also support Protección Civil's program to provide vented, wood-burning heaters, as solid fuel heaters were found to be safer than the gas heaters, even in an analysis that grouped together vented and un-vented solid fuel heaters. Electric heaters would avoid the CO risk but involve unknown fire and burn risks, and potentially higher costs. Further analysis is warranted to compare the overall relative risks and costs of electric heaters with combustion heaters. Catalytic heaters are yet another option that would appear to be safe, as they are designed for indoor use in the United States (Coleman 2004). Research would be needed to assess whether such heaters would deliver sufficient heat at a low enough cost to be attractive to residents of Ciudad Juárez.

The results of this study also suggest particularly dangerous practices that may be attractive targets for interventional programs. The population can be discouraged from sealing doors and windows when using un-vented combustion heaters and from leaving the heaters running overnight. Existing communication efforts encourage residents to open windows while operating heaters and warn them not to run the heaters while sleeping (Juárez 2004), but further efforts in this area may be justified, as well as efforts to increase the use of CO alarms in residences.

The identification of a particular manufacturer as associated with a higher risk of CO poisoning is more problematic to act on. The El Sol Corporation changed the design of its heaters. The FM-210 model, a design which used *radiantes* and which was observed to be the most prevalent of the El Sol heaters, is no longer produced (Santana 2004). Newer models, which contain an oxygen depletion sensor and an alternative heat radiation design, are believed to be safer. Thus public information encouraging the replacement of old heaters that lack oxygen depletion sensor with new heaters that have oxygen depletions sensors may be more effective than campaigns aimed encouraging or discouraging the use of particular brands of heaters. As heaters are very durable, more dangerous models may still be in use for many years if efforts to replace them are not undertaken.

While this study is far from definitive, it does tentatively identify some possible strategies for addressing the risks of CO poisoning. It is hoped that the results presented here will direct greater attention to this issue and assist with the design of follow up research projects to more rigorously identify risk factors and the development of more effective policy measures to address this public health threat.

ACKNOWLEGMENTS

This project was supported by grant A-03-01 from the EPA's Southwest Center for Environmental Research and Policy. Teresa Montoya was supported by an EPA research fellowship. Analila Rojo and Jay Graham assisted with the preparation of the baseline databases. The support and assistance of Héctor Apodaca of Protección Civil and Cindy Conroy of the Center for Environmental Research and Management is gratefully acknowledged.

REFERENCES

Apodaca, H. 2005. Proteccion Civil, Ciudad Juarez, Mexico, personal communication, January 5, 2005.

Bayer, M.J., Orlando, J., McCormick, M.A., Weiner, A., and Deckel, A.W. 1998.

"Persistent Neurological Sequelae Following Chronic Exposure to Carbon Monoxide." *Toxicology Letters,* 95(1): 78-79.

Calefactores El Sol. September 3, 2001. <http://www.calefactoreselsol.com.mx/index.php3>.

Calefactores Solmatic. November 1, 2000. <http://www.solmatic.com.mx/index1.php3>.

Diario de Juárez. 2004. "Calentadores mortales: un mal que se repite." *Diario de Juárez.* January 4, 2004, Ciudad Juarez, Mexico.

Cobb, N., and R. Etzel. 1991. "Unintentional Carbon Monoxide Related Deaths in the United States, 1979 through 1988." *JAMA,* 266(5): 659-663.

Coleman. 2004. *ProCat™ Portable Catalytic Heater with Electronic Ignition*, available online at: www.coleman.com/coleman/colemancom/detail.asp?product_id=5053A751 andcategoryid=3000, accessed December 22, 2004.

Enciclopedia de los municipios de México. September 3, 2004 <http://www.portal-juarez.com/>.

Focus Group. 2004. Comments made to P. Gurian and V. Corella-Barud by participants in a focus group held in *Colonia Siglo XXI*, Ciudad Juárez, México on September 22, 2004.

Graham, J., Gurian P, Corella-Barud V., and Avitia-Diaz Raquel. 2004. "Peri-urbanization and in-home environmental health risks: the side effects of planned and unplanned growth." *International Journal of Hygiene and Environmental Health,* 207(5): 447-454.

INEGI. 2000. *Censo de Población y Vivienda*, Instituto Nacional de Estadística, Geografía e Informática, Ciudad Juarez, Mexico.

Juárez. 2004. *Cuidado Con Los Calentadores*, Gobierno Municipal de Juárez, Ciudad Juárez, México.

Leaderer, B.P. 1982. "Air Pollutant Emissions from Kerosene Space Heaters."*Science,* 218(4577): 1113-1115.

Liu, Kai-Shen, Paz, Maria Katrina, Flessel, Peter, Waldman, Jed, and Girman, John. 2000. "Unintentional Carbon Monoxide Deaths in Calfornia from Residential and Other Nonvehicular Sources." *Archives of Environmental Health,* 55(6): 375-381

Moore, David S., and McCabe, George P. *Introduction to the Practice of Statistics*. New York: 2003.

National Weather Service Forecast Office – El Paso Area. February 17, 2004. <http://www. srh.weather.gov/elp/climat/wxclim.shtml>.

Ralston, James D. MD and Hampson, Neil B. MD 2000. "Incidence of Severe Unintentional Carbon Monoxide Poisoning Differs Across Racial/Ethnic Categories." *Public Health Reports,* 115: 46-51.

Randolph, Daley DVM, MPH, Smith, Andrew, SM, ScD, Paz-Argandona, Enrique MD MPH, Malilay, Josephine PhD, MPH, and McGeehin, Michael PhD, MSPH. 2000. "An Outbreak of Carbon Monoxide Poisoning after a Major Ice Storm in Maine." *Journal of Emergency Medicine,* 18 (1): 87-93.

Raub, James A., Mathieu-Nolf, Monique, Hampson, Neil B., and Thom, Stephen R. 2000. "Carbon monoxide poisoning – a public health perspective." *Toxicology,* 145: 1-14.

Santana, M. 2004. El Sol Corporation, personal communication, June 29, 2004.

TDH Accent on Health. November 18, 1997. <http://www.tdh.state.tx.us/ news/acc1118.htm>.

Zona Poniente Plan Parcial de Desarrollo. 2001. Ciudad Juárez, Chihuahua, Plan Director de Desarrollo Urbano.

INDEX

C

D

I

J

K

L

P

Q

R

S

T

U

V

W

X

Y

Z